Optical
Interferometry

To my father

Optical Interferometry

P. HARIHARAN

CSIRO Division of Applied Physics
Sydney, Australia

ACADEMIC PRESS
(Harcourt Brace Jovanovich, Publishers)

Sydney   Orlando   San Diego   New York
London   Toronto   Montreal   Tokyo

ACADEMIC PRESS AUSTRALIA
Centrecourt, 25-27 Paul Street North
North Ryde, N.S.W. 2113

United States Edition published by
ACADEMIC PRESS INC.
Orlando, Florida 32887

United Kingdom Edition published by
ACADEMIC PRESS, INC. (LONDON) LTD.
24/28 Oval Road, London NW1 7DX

Printed in Australia

National Library of Australia Cataloguing-in-Publication Data

Hariharan, P. (Parameswaran), 1926-
   Optical interferometry.

   Bibliography.
   Includes index.
   ISBN 0 12 325220 2.

   1. Interferometry, I. Title.

535'.4

Library of Congress Catalog Card Number: 84-72276

CONTENTS

5   Thin films   79

6   The laser as a light source   97

7   Measurements of length   117

8   The study of optical wavefronts

9   Interferometry with lasers

10   Interference spectroscopy

Preface

There has been a tremendous increase in interest in optical interferometry during the last twenty years, a period in which the subject has undergone a complete transformation. The main reason for this is, of course, the development of the laser which made available, for the first time, an intense source of coherent light. Another reason has been the increasing use of digital computers to process the data obtained from an interferometer. A third reason is the availability of single-mode optical fibres which can be used to provide an optical path several meters long, with very low noise, in an extremely small space. As a result of these advances, optical interferometry is finding a remarkably wide range of applications.

The aim of this book is to present a self-contained treatment of the subject with particular emphasis on recent developments and their implications for the future. A brief historical survey leads up to three chapters covering the classical concepts of two-beam interference, coherence and multiple-beam interference. Chapter 5 then discusses interference in thin films, antireflection coatings and interference filters.

As mentioned at the outset, lasers are now being used to an increasing extent in optical interferometry; in fact, this has led to the virtual demise of the classical mercury arc. Accordingly, Chapter 6 looks at the laser as a light source and discusses techniques for obtaining a single-frequency output and for frequency stabilization.

Five chapters then deal with applications of interferometry such as length measurement, testing optical surfaces, interference spectroscopy and

Fourier-transform spectroscopy. Emphasis is placed in these chapters on techniques which have become feasible with the development of the laser, including unequal-path interferometry, fringe-counting, heterodyne and digital interferometry, fibre-optic interferometry and nonlinear interferometry. These are followed by three chapters on holography, holographic interferometry and speckle interferometry. A final chapter on stellar interferometry describes the intensity interferometer and techniques such as stellar speckle interferometry and speckle holography. Some useful mathematical results as well as some selected topics in optics are summarized for ready reference in five appendices.

I have tried to plan this book so that it can be used by people who would like to apply interferometric techniques in their work as well as those who would like to learn more about interferometry. In the first instance, most topics are discussed at a level accessible to people with a basic knowledge of physical optics; a more detailed treatment for the serious worker then follows. Finally, the text is supplemented by a reference list of nearly 600 selected papers. Accordingly, students should find this book useful as a text, while researchers can use it as a reference work.

I am grateful to many of my colleagues for their assistance. In particular, I would like to mention W.H. ('Beattie') Steel and Philip Ciddor with whom I have had many helpful discussions. I would also like to thank Colin Chidley who was responsible for the drawings.

Acknowledgements

I would like to thank the publishers listed below, as well as the authors, for permission to reproduce figures: Optical Society of America; American Institute of Physics (Fig. 6.7); Butterworth Scientific Limited (Fig. 7.8); Cambridge University Press (Figs. 3.4–3.6); Hewlett-Packard Company (Fig. 9.1); *Japanese Journal of Applied Physics* (Fig. 8.10); *Journal de Physique et le Radium* (Fig. 11.1); *Journal of the Physical Society of Japan* (Fig. 4.4); North Holland Publishing Company (Figs. 8.6, 8.8, 8.16); Nouvelle Revue d'Optique (Fig. 15.5); Society of Photo-Optical Instrumentation Engineers (Figs. 8.15 and 9.2); Springer-Verlag (Fig. 7.1); Taylor and Francis Ltd (Fig. 15.3); The Institute of Electrical and Electronics Engineers (Fig. 9.3); The Institute of Physics (Figs. 4.8 and 14.6).

1
The development of optical interferometry

Almost everyone has come across interference phenomena such as the vivid colours in a soap bubble or in an oil slick on a wet road, or the coloured fringes seen in a thin air film enclosed between two glass plates when they are brought into contact. The latter are commonly known as 'Newtons rings' but were, in fact, first described by Boyle and, independently, by Hooke, in the latter half of the 17th century. Their discovery can be called the starting point of optical interferometry.

The development of optical interferometry extends over more than three hundred years and is closely linked with the history of wave optics. The aim of this chapter is to review briefly some of the significant stages of this development, so as to put in perspective the topics which will be discussed in more detail in this book.

1.1. The wave theory of light

We now know that the fringes seen by Boyle and Hooke were produced by the interference of light waves reflected from the two surfaces of the film. However, while Hooke did put forward a wave theory of light to explain the fringes, and this theory was expanded and put into its present form by Huyghens in 1690, it made little progress because it was opposed by Newton, who believed that a wave theory could not explain either the rectilinear propagation of light or the phenomenon of polarization. As a result, the correct explanation had to wait for almost a hundred and fifty years.

1

The barriers to the acceptance of the wave theory were first broken by Young, when, in his Bakerian lectures in 1801 and 1803, he stated the principle of interference and demonstrated that the summation of two rays of light could give rise to darkness. However, even this demonstration did not lead to immediate acceptance of the theory. Almost no one supported Young, and the discovery by Malus in 1809 that light could be polarized by reflection shook even Young's confidence, as he, like Huyghens, thought that light was propagated as longitudinal waves. The turning point came with Fresnel's brilliant memoir on diffraction in 1818, which perfected the treatment of interference, and the discovery by Arago and Fresnel that two orthogonally polarized beams could not interfere. This led Young and Fresnel to the inevitable conclusion that light waves were transverse waves. Since only longitudinal waves can propagate in a fluid, Fresnel postulated that light waves were propagated through an elastic solid pervading all matter — the 'luminiferous aether'.

1.2. Michelson's experiment

Most of the leading physicists of the 19th century supported the aether theory, even though it raised a number of questions which had no obvious answers. Thus, the phenomenon of aberration of light, which was discovered by Bradley in 1728, indicated that the aether was stationary. However, theoretical calculations by Fresnel in 1818 showed that in a medium with a refractive index n moving with a velocity v, the aether should be carried along with a velocity $v(1-1/n^2)$. Fizeau therefore carried out an experiment in 1851 with an interferometer in which the two beams traversed two columns of running water, one beam always moving with the current, while the other moved against it. This experiment, which was repeated later by Jamin and by Michelson, showed a shift of the fringes of the expected magnitude.

Based on these results, Maxwell predicted in 1880 that the movement of the earth through the aether should result in a change in the speed of light proportional to the square of the ratio of the speed of the earth to that of light. While Maxwell felt that this effect was too small to be detected experimentally, Michelson was confident that it could be observed by making use of the tremendous increase in accuracy obtained by interferometry. This led, in 1881, to Michelson's famous experiment which was designed to demonstrate the 'aether drift'. As things turned out, the null result obtained by Michelson led to the rejection of the concept of an aether and laid the foundations for the special theory of relativity [Shankland, 1973].

1.3. Measurement of the metre

Other applications of interferometry followed in rapid succession. Thus, in 1896, Michelson carried out the first measurement of the length of the Pt–Ir bar which was the international prototype of the metre in terms of the wavelength of the red cadmium radiation. While the idea of the wavelength of a monochromatic source as a natural standard of length had been suggested much earlier by Babinet and by Fizeau, it was Michelson's work which demonstrated its feasibility and led, in 1960, to the redefinition of the metre in terms of the wavelength of the orange radiation of ^{86}Kr.

Another major field of application of interferometry was opened up by Twyman in 1916, when he used a modified Michelson interferometer to test optical components. This interferometer was, in turn, adapted by Linnik in 1933 to permit microscopic examination of reflecting surfaces. Along with this, interferometry became a valuable tool in studies of fluid flow and combustion.

1.4. Coherence

Studies by Michelson also revealed the connection between the visibility of the fringes in his interferometer and the dimensions and spectral purity of the source. Since any thermal source can be considered as made up of many elementary radiators (atoms) which are not synchronized, the interference pattern with such a source is obtained by adding the intensities in the interference patterns formed by these incoherent elementary radiators. The gradual transition from incoherent to coherent illumination with a thermal source was demonstrated as far back as 1869 by Verdet, who showed that light from two pinholes illuminated by the sun produced an interference pattern on a screen if their separation was less than 0.05 mm. However, the first quantitative concepts of coherence were formulated by von Laue only in 1907, and it was again only after a long delay that the foundations of modern coherence theory were laid in three papers by van Cittert [1934, 1939] and Zernike [1938]. These were developed in more detail twenty years later by Hopkins [1951, 1953] and by Wolf [1954, 1955]. The discovery by Hanbury Brown and Twiss [1954] of intensity correlation effects (fourth-order correlation) led to the formulation of a general description of higher-order coherence effects by Mandel and Wolf [1965].

1.5. Interference filters

An experimental observation which was made by Fraunhofer in 1817, and by Taylor in 1891, was that a glass surface with a thin layer of tarnish had a lower reflectance than a fresh glass surface. This observation was exploited on a large scale forty years later, when advances in vacuum technology made it possible for the first time to produce a range of dielectric coatings on glass surfaces by evaporation. Very effective antireflection coatings were produced by Strong [1936]; these were followed shortly afterwards by the first interference filters, opening up the whole area now known as thin film optics.

1.6. Interference spectroscopy

Towards the end of the 19th century, interference techniques found their way into high-resolution spectroscopy, with the development of instruments such as the Fabry-Perot étalon, the Michelson echelon and the Lummer-Gehrcke plate. Of these, the most widely used, initially, was the Lummer-Gehrcke plate. However, as improved multilayer dielectric coatings, which gave high values of reflectance with negligible losses, became available, the greater light-gathering power (or étendue) of the Fabry-Perot interferometer led to its rapidly replacing the others.

At this stage, a completely new approach was opened up by the development of Fourier transform spectroscopy. The origins of this technique can be traced back to 1862, when Fizeau studied the effect of the separation of the plates on Newton's rings formed with sodium light. He found that the rings almost disappeared at a separation corresponding to the passage of 490 fringes, but they reached maximum contrast once again when 980 fringes had passed, showing that the sodium line was a doublet. This method was taken a step further by Michelson in 1891 when he plotted the visibility of the fringes as a function of the optical path difference for a number of spectral lines. All of these, with the exception of the red cadmium line exhibited a series of maxima and minima, indicating that they consisted of more than one component. A similar technique was also applied to far infrared spectroscopy ($\lambda = 100$ to $300 \ \mu m$) by Rubens and Wood in 1911. However, its systematic development started with Fellgett [1951] who was the first to obtain a spectrum from a numerically Fourier-transformed interferogram and demonstrate the advantages of Fourier-transform spectroscopy. Subsequent improvements have brought this technique to a point where it reigns supreme in the infrared and is preferable to conventional dispersive instruments in the visible region when complex spectra are to be mapped with the highest possible resolution.

1.7. The development of the laser

Throughout the first half of the twentieth century, the most commonly used light source for interferometry was a pinhole illuminated by a mercury arc through a filter which isolated the green line (λ = 546 nm). Such a source only gives light with limited spatial and temporal coherence and, in addition, has a very low intensity. A revolution in interferometry began with the development of the laser, which made available, for the first time, an intense source of light with a remarkably high degree of spatial and temporal coherence.

The origin of the laser can be traced back to 1917, when Einstein pointed out that atoms in a higher energy state, which normally radiate spontaneously, could also be stimulated to emit and revert to a lower energy state when irradiated by a wave of the correct frequency. The most remarkable feature of this process was that the emitted photon had the same frequency, polarization and phase as the stimulating wave and propagated in the same direction. Schawlow and Townes [1958] were the first to show that amplification by stimulated emission was possible in the visible region and that a simple resonator consisting of two mirrors could be used for mode selection. The first practical laser was the pulsed ruby laser [Maiman, 1960]. Continuous laser action was achieved soon afterwards with the helium-neon laser, first in the infrared [Javan et al., 1961] and then in the visible region [White and Rigden, 1962]. The high degree of directionality and coherence of laser light were also verified by Collins et al., [1960]. Lasers have removed most of the limitations of interferometry imposed by thermal sources and have made possible a dramatic change in techniques.

1.8. Holography

One such technique is holography, which is a completely new method of imaging based on optical interference. It was first demonstrated by Gabor in 1948, though its roots can be traced back to work by Wolfke [1920] and Bragg [1939] on the determination of the structure of a crystal from its X-ray diffraction pattern. However, for the next fifteen years, very little work was done on holography until the invention of the laser and the development of the off-axis reference-beam method made it possible to record holograms of diffusely reflecting objects having appreciable depth [Leith and Upatnieks, 1964]. This triggered off a flood of publications describing applications of the new technique.

Perhaps the most significant of these is holographic interferometry, which was discovered independently, and almost simultaneously, by several groups [Brooks et al., 1965; Burch, 1965; Collier, Doherty and Pennington, 1965; Haines and Hildebrand, 1965; Powell and Stetson, 1965]. Holographic interferometry made it possible, for the first time, to map the displacements of a rough surface with an accuracy of a few nanometres, and even to make interferometric comparisons of two stored wavefronts that existed at different times.

New techniques such as computer-generated holograms [Lohmann and Paris, 1967] as well as the development of new recording materials, such as photothermoplastics and photorefractive crystals, have opened up a number of interesting possibilities for holography.

1.9. Speckle

A phenomenon which attracted immediate attention following the development of continuous-wave lasers was the granular appearance of a diffusing surface illuminated by the laser beam. This phenomenon, which was named 'speckle', was shown to be due to the interference of the completely coherent diffracted beams from individual elements on the diffusing surface [Rigden and Gordon, 1962]. Its characteristics were analyzed in detail in a number of papers before it was realized that such phenomena had, in fact, been studied much earlier.

Thus, Newton, and later on, Quetélet, observed the coloured rings formed by the interference of scattered light when a back-silvered mirror was illuminated by a point source of white light. Subsequently, in 1877, Exner found that the rings produced when monochromatic light from a point source is diffracted by a glass plate covered with small particles exhibited a granular structure. He concluded that the superposition of a large number of identical wavetrains with random phase differences must lead to sharp fluctuations in intensity. Detailed theoretical studies of these fluctuations were made by von Laue in 1914 and 1916, which anticipated many of the findings of more recent work [see Hariharan, 1972].

Laser speckle was regarded, initially, as a nuisance. However, it was soon realized that speckle could be used as a carrier of information. As a result, speckle has found a number of applications, of which the best known are in studies of surface displacements and vibrations [Archbold et al., 1969; Leendertz, 1970], where speckle interferometry is a valuable supplement to holographic interferometry.

1.10. Stellar interferometry

The idea of using interference to measure the angular dimensions of a star goes back to 1868, when Fizeau proposed an arrangement consisting of two slits placed in front of a telescope; their separation when the interference fringes crossing the image of the star disappeared could then be used to calculate the angle subtended by the star. Unfortunately, this experiment failed because the aperture of the telescope was not large enough.

However, Fizeau's proposal was taken up again by Michelson, who applied it successfully to measure the diameters of Jupiter's satellites in 1890. This success was followed in 1921 by measurements of the diameter of Betelgeuse and six other stars, using a 6 m stellar interferometer mounted on the 2.5 m (100 inch) telescope at Mt Wilson.

An attempt by Michelson to extend this technique to longer baselines failed because of problems of mechanical stability and atmospheric turbulence, and the next advance came from a completely different approach, namely the idea of measuring the degree of correlation of the intensity fluctuations in the source [Hanbury Brown and Twiss, 1954]. This new method was used in the intensity interferometer [Hanbury Brown et al., 1964] to make measurements on 32 stars with angular diameters down to 0.0004 second of arc.

Other developments in this field which followed have been the techniques of stellar speckle interferometry [Labeyrie, 1970] and speckle holography [Liu and Lohmann, 1973; Bates et al., 1973] which have made it possible to determine the structure of many close groupings of stars; these confirm that this field still remains quite active.

1.11. Digital techniques

Another development which has revolutionized interferometry has been the increasing use of electronics. This trend started with the use of photoelectric detectors with Fabry-Perot interferometers, but soon extended to such applications as fringe-counting interferometers for length measurements. Digital computers were first used in Fourier spectroscopy, and have made it an extremely powerful tool. Digital systems have also made possible direct measurements of the optical path difference at an array of points covering an interference pattern.

1.12. Heterodyne techniques

The availability of a highly coherent source in the form of the laser was followed by the observation of beats when the beams from two lasers operating at slightly different frequencies were mixed at a photodetector [Javan et al., 1962]. This led to the development of a range of heterodyne techniques to replace traditional methods of interpolation. Since measurements of either the frequency or the phase of a beat can be made with very high precision, such techniques have revolutionized length interferometry. Another result was the extension of frequency measurements to the optical region and the redefinition of the metre in terms of the speed of light.

1.13. Fibre interferometers

Another major advance followed the use of single-mode fibres to build analogues of conventional two-beam interferometers. Such interferometers have the advantage that very long optical paths can be accommodated in a small space. They have found many applications since the optical path through a length of such a fibre changes with pressure or temperature and it can be used as a sensor. Extremely high sensitivity is possible with fibre interferometers since they have very low noise and sophisticated detection techniques can be used with them.

1.14. Future directions

It is always an interesting, though risky, exercise to try to guess what are going to be the directions of future development in any field. In the case of optical interferometry, some trends are obvious, such as the growing use of lasers and digital signal processing to increase the scope, speed and precision of measurements. With the replacement of the human eye by electronic detectors, the advantages of using infrared wavelengths for interferometry are being exploited to an increasing extent. Other interesting possibilities have been opened up by the application of single-mode fibres and nonlinear crystals to build new types of interferometers.

2
Two-beam interference

A beam of light is actually a propagating electromagnetic wave. If, for simplicity, we assume that we are dealing with a linearly polarized plane wave propagating in a vacuum in the z direction, the electric field E at any point can be represented by a sinusoidal function of distance and time,

$$E = a \cos [2\pi\nu (t - z/c)] \tag{2.1}$$

where a is the amplitude, ν the frequency, and c the velocity of propagation of the wave. If T is the period of the vibration and ω its circular frequency,

$$T = 1/\nu = 2\pi/\omega \tag{2.2}$$

The wavelength λ is given by the relation

$$\lambda = cT = c/\nu \tag{2.3}$$

while the propagation constant of the wave is

$$k = 2\pi/\lambda \tag{2.4}$$

However, the wavelength also depends on the velocity of propagation of the wave, and hence, on the refractive index of the medium. If the medium is changed, the wavelength changes even though the frequency is the same. In a medium with a refractive index n, in which the light-wave propagates with a velocity

$$v = c/n \tag{2.5}$$

the wavelength of the same radiation would be

$$\lambda_n = vT$$
$$= v\lambda/c = \lambda/n \tag{2.6}$$

2.1 Complex representation of monochromatic light waves

Equation (2.1) can also be written as

$$E = \mathrm{Re}\ \{a \exp [\mathrm{i}2\pi\nu(t - z/v)]\}, \tag{2.7}$$

where Re {} represents the real part of the expression within the braces, and $i = (-1)^{\frac{1}{2}}$. This has the advantage that the right hand side can now be expressed as the product of spatially varying and temporally varying factors, so that

$$E = \mathrm{Re}\ \{a \exp (-\mathrm{i}2\pi\nu z/v) \exp (\mathrm{i}2\pi\nu t)\}$$
$$= \mathrm{Re}\ \{a \exp (-\mathrm{i}\phi) \exp (\mathrm{i}2\pi\nu t)\} \tag{2.8}$$

where

$$\phi = 2\pi\nu z/v$$
$$= 2\pi n z/\lambda \tag{2.9}$$

The product $p = nz$ is the optical path between the origin and the point z, and ϕ is the corresponding phase difference.

If we assume that all operations on E are linear, it is simpler to use the complex function

$$E = a \exp (-\mathrm{i}\phi) \exp (\mathrm{i}2\pi\nu t) \tag{2.10}$$

and take the real part at the end of the calculation. We can then rewrite Eq. (2.10) as

$$E = A \exp (\mathrm{i}2\pi\nu t) \tag{2.11}$$

where

$$A = a \exp(-i\phi) \tag{2.12}$$

is known as the complex amplitude of the vibration.

Because of the extremely high frequencies of visible light waves ($\nu \approx 6 \times 10^{14}$ Hz for $\lambda = 0.5\mu$m) direct observations of the electric field are not normally possible. The only measurable quantity is the intensity, which is the time average of the amount of energy which, in unit time, crosses a unit area normal to the direction of the energy flow. This is proportional to the time average of the square of the electric field.

$$< E^2 > = \lim_{T \to \infty} \frac{1}{2T} \int_{-T}^{T} E^2 \, dt \tag{2.13}$$

From Eqs. (2.1), (2.2) and (2.9) we have

$$< E^2 > = \lim_{T \to \infty} \frac{1}{2T} \int_{-T}^{T} a^2 \cos^2(\omega t - \phi) \, dt$$
$$= a^2/2. \tag{2.14}$$

Since we are not interested in the absolute value of the intensity, but only in relative values over a specified region, we can ignore this factor of 1/2, as well as any other factors of proportionality, and define the optical intensity as

$$I = a^2 = |A|^2 \tag{2.15}$$

2.2. Interference of two monochromatic waves

If two monochromatic waves propagating in the same direction and polarized in the same plane are superposed at a point P, the total electric field at this point is

$$E = E_1 + E_2 \tag{2.16}$$

where E_1 and E_2 are the electric fields due to the two waves. If the two waves have the same frequency, the intensity at this point is, from Eq. (2.4)

$$I = |A_1 + A_2|^2 \tag{2.17}$$

where $A_1 = a_1 \exp(-i\phi_1)$ and $A_2 = a_2 \exp(-i\phi_2)$ are the complex amplitudes of the two waves. Accordingly,

$$I = A_1^2 + A_2^2 + A_1 A_2^* + A_1^* A_2$$
$$= I_1 + I_2 + 2(I_1 I_2)^{\frac{1}{2}} \cos \Delta\phi \qquad (2.18)$$

where I_1 and I_2 are the intensities at P due to the two waves acting separately, and $\Delta\phi = \phi_1 - \phi_2$ is the phase difference between them.

If Δp is the corresponding difference in the optical paths given by Eq. (2.9), the order of interference is $N = \Delta p/\lambda$. The intensity has its maximum value I_{max} when

$$N = m, \quad \Delta p = m\lambda, \quad \Delta\phi = 2m\pi \qquad (2.19)$$

where m is an integer, and its minimum value I_{min} when

$$N = (2m+1)/2, \quad \Delta p = (2m+1)\lambda/2, \quad \Delta\phi = (2m+1)\pi \qquad (2.20)$$

A convenient measure of the contrast of the interference phenomenon is the visibility which is defined by the relation

$$\mathcal{V} = (I_{max} - I_{min})/(I_{max} + I_{min}) \qquad (2.21)$$

In the present case, it follows from Eqs (2.18–2.21) that

$$\mathcal{V} = 2 (I_1 I_2)^{\frac{1}{2}}/(I_1 + I_2) \qquad (2.22)$$

However, light from a thermal source, even when it comprises only a single spectral line, is not strictly monochromatic. Both the amplitude and the phase exhibit very rapid, random fluctuations, and, in the case of beams from different sources, these fluctuations are not correlated. Interference effects cannot therefore be observed between beams derived from two thermal sources. To simplify matters we shall assume, at this stage, that the correlation between the fluctuations in the beams is complete, in which case they are said to be completely coherent. The elementary theory developed earlier for monochromatic light is then adequate to describe the effects observed.

We shall consider, in the first instance, interference between two beams; the phenomena observed can then be classified according to the method used to obtain these beams.

2.3 Wavefront division

One way to obtain two beams from a single source is to use two separate portions of the original wavefront. These are then superposed to produce interference. This method is known as wavefront division.

A simple optical system for this purpose is the arrangement known as Fresnel's mirrors. In this, as shown in Fig. 2.1, light from a point source S is reflected at two mirrors M_1 and M_2 which make a small angle ϵ with each other. Interference fringes are observed in the region where the two reflected beams overlap. Interference can be considered to take place between the light from the two secondary sources S_1 and S_2, the images of S in the mirrors M_1 and M_2. If d is the distance of S from A, we have

$$S_1 S_2 = 2d\epsilon \qquad (2.23)$$

and the path difference between the waves from these two virtual sources to a point $P(x,y)$ on the screen in the neighbourhood of O is, to a first approximation,

$$p = 2d\epsilon x/(D+d) \qquad (2.24)$$

Since successive maxima or minima in the interference pattern correspond to a change in the path difference

$$\Delta p = \lambda \qquad (2.25)$$

the fringes approximate to equidistant straight lines running parallel to the Y axis; their separation Δx is given by the relation

$$\Delta x = \lambda(D+d)/2d\epsilon \qquad (2.26)$$

An alternative setup is Lloyd's mirror. In this arrangement, a point source illuminates a mirror at near grazing incidence and interference takes place between the light from the two sources constituted by the primary source and its image in the mirror.

2.4 Amplitude division

The other method by which two beams can be obtained from a single source is by division of the amplitude over the same section of the wavefront. One way is to use a surface which reflects part of the incident light and transmits part of it.

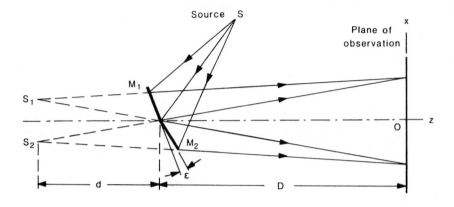

Fig. 2.1. Optical system used to produce interference by wavefront division (Fresnel's mirrors).

2.4.1 Interference in a plane-parallel plate

Consider a transparent plane-parallel plate illuminated, as shown in Fig. 2.2, by a point source of monochromatic light. Any point P on the same side of the plate as the source receives two beams of nearly equal amplitude from it, one reflected from the upper surface of the plate and the other from its lower surface. We can see from considerations of symmetry that the interference fringes observed in a plane parallel to the plate are circles with their centre at $0'$, the point where this plane intersects $S0$, the normal to the plate.

A case of particular interest is when the plane of observation is at infinity. This is the situation when the fringes are observed in the back focal plane of a lens, as shown in Fig. 2.3. In this case the two interfering rays AL and CL' are parallel and are derived from the same incident ray. Let the thickness of the plate be d and its refractive index n_2, while that of the medium on both sides of it is n_1.

If, then, θ_1 and θ_2 are respectively the angles of incidence and refraction at the upper surface, we have

$$AB = BC = d/\cos \theta_2 \qquad (2.27)$$

$$AC = 2d \tan \theta_2 \qquad (2.28)$$

and

$$AD = AC \sin \theta_1 = 2d \tan \theta_2 \sin \theta_1 \qquad (2.29)$$

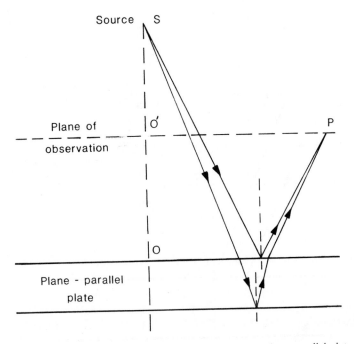

Fig. 2.2. Formation of interference fringes by reflection in a plane-parallel plate.

Accordingly, the optical path difference between the two rays should be

$$\Delta p = n_2 (AB + BC) - n_1 AD$$
$$= 2n_2 d \cos \theta \qquad (2.30)$$

However, we also have to take into account an additional phase shift of π introduced by reflection at one of the surfaces (see Appendix A.4); the optical path difference between the interfering wavefronts is, therefore, actually

$$\Delta p = 2n_2 d \cos \theta \pm \lambda/2 \qquad (2.31)$$

A bright fringe corresponds to the condition

$$2 n_2 d \cos \theta_2 \pm \lambda/2 = m\lambda \qquad (2.32)$$

where m is an integer, while a dark fringe corresponds to the condition

$$2 n_2 d \cos \theta_2 \pm \lambda/2 = (2m + 1) \lambda/2 \qquad (2.33)$$

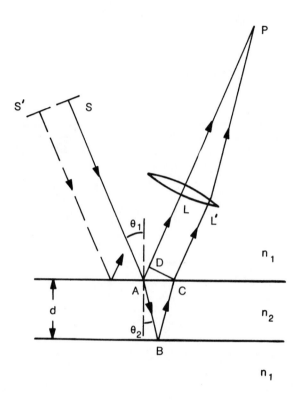

Fig. 2.3. Formation of fringes of equal inclination by reflection in a plane-parallel plate.

Equation (2.31) shows that for a given value of d the phase difference between the wavefronts depends only on the angle θ_2. This makes it possible to use an extended monochromatic source instead of a point source. The interference fringes produced in the back focal plane of L by any other point S' on an extended source are identical with those produced by S, so that their visibility is unaffected. The fringes are circles centred on the normal to the plate (Haidinger's rings) and are called fringes of equal inclination.

For near-normal incidence θ_2 is small and Eq. (2.33) can be written as

$$2\,n_2d\,(1\,-\,\theta_2{}^2/2)\,=\,m\lambda \tag{2.34}$$

If there is an intensity minimum (or maximum) at the centre, the radii of the dark (or bright) rings are proportional to the square roots of consecutive integers.

Similar phenomena can also be observed in transmission. In this case the directly transmitted beam interferes with the beam formed by two internal reflections. Since the net phase shift introduced by the reflections at the two surfaces of the plate is either zero or 2π, the optical path difference between the beams is

$$\Delta p = 2nd \cos \theta_2 \qquad (2.35)$$

The fringes are, therefore, complementary to those seen by reflection. However, since the relative amplitudes of the two beams are usually very different (1.00 and 0.05 for a glass plate in air), the visibility of the fringes is low.

2.4.2 Interference in a plate of varying thickness: Fizeau fringes

We assume that the plate has only small variations of thickness and that it is illuminated at near-normal incidence with a collimated beam of monochromatic light as shown in Fig. 2.4. The lens forms an image of the plate at P. Consider two parallel incident rays which follow the paths $S_1 ABCL_1 P$ and $S_2 CL_2 P$. Since the optical paths $CL_1 P$ and $CL_2 P$ are equal, it is sufficient to consider the difference in the optical paths between C and the source. In addition, since A and C are very close together, we can neglect the variation in thickness between these points and assume that the plate has a well defined thickness d over this region. The optical path difference is then, from Eq. (2.31),

$$\Delta p = 2n_2 d \cos \theta_2 \pm \lambda/2 \qquad (2.36)$$

which, when θ_2 is small reduces to

$$\Delta p = 2n_2 d \pm \lambda/2 \qquad (2.37)$$

The fringes then correspond to contours of equal thickness and are known as Fizeau fringes.

In this case also a complementary fringe pattern of low contrast is seen by transmission.

2.4.3 Interference in a thin film

Fringes of equal thickness can be produced in a thin film even without the use of collimated light. With an extended source, it follows from Eq. (2.36) the optical path difference varies with the angle θ_2 as well as with the thickness d. However, for a thin film (for which d is very small), if the

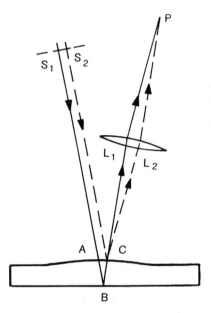

Fig. 2.4.   Formation of fringes of equal thickness (Fizeau fringes) by reflection.

interference fringes are viewed with an optical system (or with the eye) focused on the film, the pupil limits the range of values of θ_2 for any point on the film, so that, for near-normal incidence ($\cos \theta_2 \approx 1$), the variation of the optical path difference over this range of angles is negligible. Under these conditions, the fringes are effectively fringes of equal thickness.

2.5. Localization of fringes

So far, we have studied the phenomenon of interference under conditions which permit a relatively simple analysis. The principal restriction imposed was the use of a point source or a collimated beam of monochromatic light. We shall now examine the consequences of relaxing this restriction on the size of the source.

2.5.1 Nonlocalized fringes

Consider the optical system shown in Fig. 2.5, which produces two images S_1 and S_2 of the point source S. Interference fringes are produced where the beams from these two secondary sources overlap. At a point P, the optical path difference between the beams is

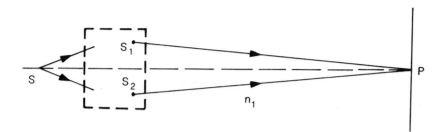

Fig. 2.5. Formation of nonlocalized fringes with a point source.

$$\Delta p = n_1(S_2P - S_1P) \qquad (2.38)$$

where n_1 is the refractive index of the surrounding medium. The value of Δp depends on the position of P; interference fringes corresponding to the loci of points for which Δp is constant are therefore formed over the plane containing P. Since this plane can take any position within the region of superposition of the beams, the fringes are said to be nonlocalized. Such fringes are always obtained with a point source. If, in addition, it is a monochromatic source, their visibility depends only on the relative intensities of the two images S_1 and S_2.

2.5.2 Localized fringes

Let us now consider what happens when, as shown in Fig. 2.6, the point source S is replaced by an extended source. Such a source can be considered as an array of independent point sources, each of which produces a separate interference pattern. If the path differences at P are not the same for all these point sources, these elementary fringe patterns do not coincide, and there is a reduction in the visibility of the fringes. Since the reduction in visibility depends on the position of P, there is, in general, a position of the plane of observation for which the visibility of the fringes is a maximum; the fringes are then said to be localized at this plane.

Localization of the fringes is a normal consequence of the use of an extended source. We can study the phenomenon in more detail by examining how the path difference at P for a generalized point S' in the source varies with its position. For this it is convenient to interchange the source plane and the plane of observation and consider the situation if we were to have a point source at P. This point source would then give rise to two spherical wavefronts W_1 and W_2 passing through S which are normal to SI_1 and SI_2, respectively. The path difference between these wavefronts at S is

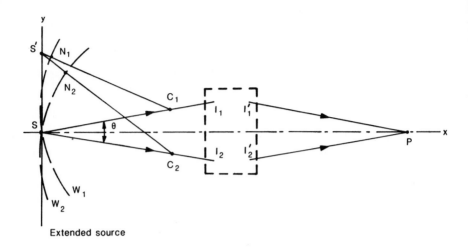

Extended source

Fig. 2.6. Localization of fringes with an extended source.

obviously given by Eq. (2.38). If now, the normals from S' to W_1 and W_2 are $S'N_1$ and $S'N_2$ respectively, these normals represent the additional paths traversed by these wavefronts before they reach S'. Accordingly, the path difference between the two wavefronts from S' at P is

$$\Delta p' = \Delta p + \Delta' p \qquad (2.39)$$

where $\Delta' p = n_1 (S'N_2 - S'N_1)$.

To calculate $\Delta' p$ it is convenient to choose the bisector of the angle $I_1 S I_2$ and the normal to it as the coordinate axes SX, SY. The coordinates of C_1 and C_2, the centres of the wavefronts W_1 and W_2, are then $[R_1 \cos (\theta/2), R_1 \sin (\theta/2)]$ and $[R_2 \cos (\theta/2), -R_2 \sin (\theta/2)]$, respectively, where $R_1 = C_1 S$ and $R_2 = C_2 S$ are the radii of curvature of W_1 and W_2. Accordingly, since S' has the coordinates $(0, y)$, and $l_1 = S'N_1$, $l_2 = S'N_2$

$$C_1 S' = R_1 + l_1$$
$$= \{R_1^2 \cos^2 (\theta/2)` + [y - R_1 \sin (\theta/2)]^2\}^{1/2} \qquad (2.40)$$

and

$$C_2 S' = R_2 + l_{2'}$$
$$= \{R_2^2 \cos^2 (\theta/2) + [y + R_2 \sin (\theta/2)]^2\}^{1/2} \qquad (2.41)$$

If the dimensions of the source are small compared to R_1 and R_2, terms involving powers of (y/R_1) and (y/R_2) higher than the second can be neglected, so that we have

$$l_1 = -y \sin(\theta/2) + (y^2/2R_1) \cos^2(\theta/2) \qquad (2.42)$$

and

$$l_2 = y \sin(\theta/2) + (y^2/2R_2) \cos^2(\theta/2) \qquad (2.43)$$

so that

$$\Delta'p = n_1 (l_2 - l_1)$$
$$= n_1 \{2y \sin(\theta/2) + (y^2/2) [(1/R_2) - (1/R_1)] \cos^2(\theta/2)\} \qquad (2.44)$$

or, since θ is usually small,

$$\Delta'p \approx n_1 \{y\theta + (1/2) [(1/R_2) - (1/R_1)] y^2\} \qquad (2.45)$$

If the source is small enough for the term involving y^2 in Eq. (2.45) to be neglected,

$$\Delta'p = n_1 y\theta \qquad (2.46)$$

The fringes then have maximum visibility at P when $\Delta'p = 0$, that is to say, when $\theta = 0$. This occurs when SI_1 and SI_2 coincide or, in other words, the two rays $I_1'P$ and $I_2'P$ which intersect at P originate in the same ray coming from S. It follows that the interference fringes are localized on a surface which is the locus of the points of intersection of pairs of rays, each of which is derived from a single ray leaving the source.

In the vicinity of this surface of localization $y\theta$ is negligible, and Eq. (2.45) reduces to

$$\Delta'p = n_1 [(1/R_2) - (1/R_1)] y^2/2 \qquad (2.47)$$

It is apparent from Eq. (2.47) that the visibility of the fringes in the plane of localization decreases as the size of the source increases. For good visibility $\Delta'p$ must be less than a small fraction of the wavelength (say, $\lambda/4$). The maximum dimensions of the source are then given by the condition that

$$n_1 [(1/R_2) - (1/R_1)] y^2/2 \leq \lambda/4 \qquad (2.48)$$

or

$$y \leq (\lambda R_1 R_2/2n_1|R_1 - R_2|)^{1/2} \qquad (2.49)$$

We shall apply these results to some specific cases in the next two sections.

2.5.3 Fringes in a plane-parallel plate

In the case of a plane-parallel plate (see Section 2.4.1), when the fringes are observed in the back focal plane of a lens, $\theta = 0$, and R_1 and R_2 are infinite for all positions of S'. Accordingly, the fringes are localized at infinity, and their visibility does not decrease with an extended source.

2.5.4 Fringes in a thin film

Consider a thin layer of refractive index n_2 contained between two surfaces of a material with a refractive index n_1 which, as shown in Fig. 2.7, make a small angle ϵ with each other. A ray from a point source S gives rise to two reflected rays, one reflected from the upper surface and the other from the lower surface, which meet at P. It follows that with an extended source centred on S, the fringes of equal thickness formed with such a wedge, which are straight lines running at right angles to its principal section, are localized at P. When, as in Fig. 2.7 (a), the points B and O are on the same side of the normal at A, the point P lies on the same side of the wedge as S. However, when, as in Fig. 2.7 (b), the points B and O are on opposite sides of the normal at A, the point P lies on the other side of the wedge.

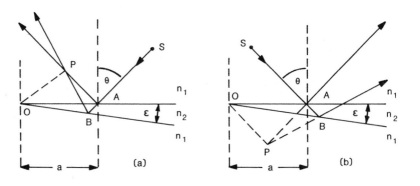

Fig. 2.7. Localization of fringes in a wedged film.

A detailed analysis [Born & Wolf, 1980] shows that in both cases

$$AP \approx (an_1^2 \sin \theta \cos^2 \theta)/(n_2^2 - n_1^2 \sin^2 \theta) \qquad (2.50)$$

where θ is the angle of incidence and a is the distance of the point of incidence on the upper surface from the apex of the wedge. For an air film between thin glass plates, the effects of refraction at the upper surface may be neglected and we can set $n_1 = n_2 = 1$. We then have

$$AP \approx a \sin \theta \qquad (2.51)$$

and the angle OPA is very nearly a right angle.

The coordinates of P can be obtained from Fig. 2.8; they are

$$x = u \cos^2 \theta - v \sin \theta \cos \theta \qquad (2.52)$$

and

$$y = u \cos \theta \sin \theta - v \sin^2 \theta \qquad (2.53)$$

where (u,v) are the coordinates of S. It follows, therefore, that the locus of P as A, the point of incidence, moves across the surface is a circle with its

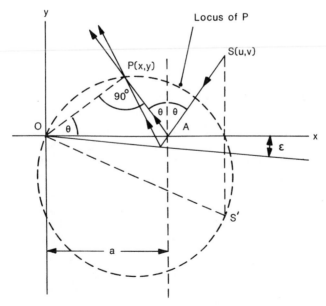

Fig. 2.8. Locus of the point of localization with changing angle of incidence.

centre at $(u/2, -v/2)$ or, in other words, a circle with OS' as a diameter, where S' is the reflected image of S in the upper surface of the wedge. In the limiting case when S is at infinity, corresponding to illumination with a parallel beam, the locus of P is a plane passing through the vertex of the wedge. For normal incidence, this plane of localization coincides with the front surface of the wedge.

To determine, in this case, the maximum size of the source that does not lead to an appreciable drop in the visibility of the fringes, we imagine a point source placed at P and evaluate R_1 and R_2 the radii of curvature of the wavefronts W_1 and W_2 reaching S after reflection at the upper and lower surfaces of the wedge. This calculation [see Born & Wolf, 1980] shows that, in general, W_2 has different radii in the principal section and at right angles to it. However, for near-normal incidence this difference can be neglected and we have $|R_2 - R_1| = 2dn_1/n_2$, where d is the thickness of the film at the point of incidence. In addition, if the source is at a relatively great distance from the film compared to its thickness, Eq. (2.49) can be rewritten in terms of α the angular radius of the source, as seen from P, as

$$\alpha = (1/R_1) \, (\lambda \, R_1 R_2 / 2n_1 \, |R_1 - R_2|)^{\frac{1}{2}}$$
$$\approx (\lambda n_2 / 4n_1 d)^{\frac{1}{2}} \qquad (2.54)$$

since $R_2/R_1 \approx 1$

2.6. Two-beam interferometers

For many applications it is desirable to have an optical arrangement in which the two interfering beams travel along separate paths before they are recombined. This requirement has led to the development of a large number of interferometers for specific purposes. However, apart from the Rayleigh interferometer and its variations, which make use of wavefront division, most are based on division of amplitude.

Three methods have been described for division of amplitude. The first uses a partially reflecting film of a metal or dielectric, commonly called a beam splitter. Another uses a birefringent element to produce two orthogonally polarized beams. However, for these to interfere they must be derived from a polarized beam and then brought once again into the same state of polarization. In a third method, a grating or a scatter plate (a surface whose amplitude transmittance varies in a random manner) produces one or more diffracted beams as well as a reflected or transmitted beam.

Two-beam interferometers can also have different configurations. These are typified by the Michelson, Mach–Zehnder and Sagnac interferometers.

2.7. The Michelson interferometer

In this instrument, as shown in Fig. 2.9, light from a source S is divided at a semi-reflecting coating on one surface of a plane-parallel glass plate B into two beams with nearly equal amplitudes. These are reflected back at two plane mirrors M_1, M_2, and return to B where they are recombined and emerge at O.

The optical path difference p between the two arms of the interferometer is given by the difference of two summations, taken respectively over the two arms, of the products of the thickness of each medium traversed and its refractive index, so that

$$p = \sum_2 nd - \sum_1 nd \qquad (2.55)$$

The optical path difference defined by Eq. (2.55) is normally dependent on the wavelength, since all glasses have a finite dispersion $dn/d\lambda$ which is also a function of wavelength. If the optical path difference p is to be strictly independent of wavelength, it is necessary for both the arms to contain the same thickness of glass having the same dispersion. Since in the Michelson interferometer one beam traverses the beam splitter B only once, while the other traverses it three times, a compensating plate C of the same material and having the same thickness as B is introduced in the first beam.

Reflection at the beam splitter B produces an image of the mirror M_2 at M_2'. The interference pattern observed is therefore the same as that produced in a layer of air bounded by the mirror M_1 and M_2', the virtual image of M_2.

2.7.1 Nonlocalized fringes

A monochromatic point source S located at a finite distance gives rise to two virtual point sources, S_1 and S_2, which are the images of S reflected in M_1 and M_2. Nonlocalized fringes are then observed on a screen placed at O. If, as shown in Fig. 2.10(a) the line S_1S_2 is parallel to AO, the fringes are circular; if, as in Fig. 2.10(b), S_1S_2 is at right angles to AO, they are parallel straight lines.

2.7.2 Fringes of equal inclination

With an extended monochromatic source, circular fringes of equal inclination are formed when M_1 and M_2' are parallel. These are localized at infinity and can be observed at O either with a telescope or directly with the eye relaxed. If M_1 is moved closer to M_2', the fringes contract towards the centre

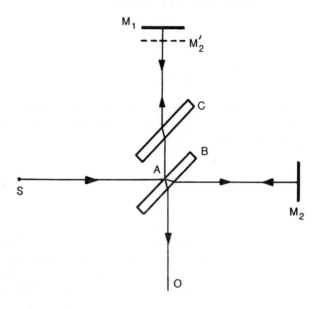

Fig. 2.9. The Michelson interferometer.

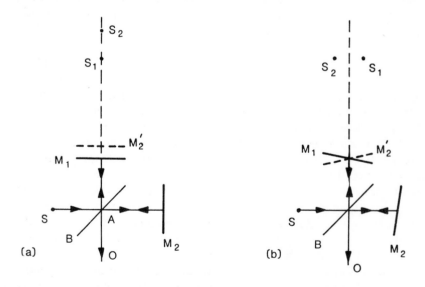

Fig. 2.10. Formation of nonlocalized fringes in a Michelson interferometer.

of the pattern where, one after the other, they vanish; at the same time, the scale of the pattern increases until, when M_1 and M'_2 coincide, a uniform field is obtained.

2.7.3 Fringes of equal thickness

If M_1 and M'_2 make a small angle with each other and their separation is very small, fringes of equal thickness can be seen with an extended source. These are equidistant straight lines parallel to the apex of the wedge and are localized at $M_1M'_2$.

2.8. The Mach–Zehnder interferometer

As shown in Fig. 2.11, the Mach–Zehnder interferometer contains two beam splitters B_1, B_2, and two mirrors M_1, M_2. Light from a source S is divided at the semireflecting surface of B_1 into two beams, which, after reflection at the mirrors M_1, M_2 are recombined at the semireflecting surface of B_2. Usually B_1, B_2 and M_1, M_2 are adjusted so that they are approximately parallel and the paths traversed by the beams form a rectangle or a parallelogram.

We assume that the interferometer is illuminated with a collimated beam giving rise to two plane wavefronts, say W_1 and W_2, in the two arms. Let W'_2 be the image of the plane W_2 in the beam splitter B_2. The phase difference between W_1 and W'_2 at a point P on W_1 is then

$$\Delta\phi = knd \tag{2.56}$$

where d is the separation of W_1 and W'_2 at P, and n is the refractive index of the medium between them. If W_1 and W'_2 make a small angle with each other, a nonlocalized interference pattern is seen, consisting of equispaced straight fringes parallel to the line of intersection of W_1 and W'_2.

A similar fringe pattern is also obtained with an extended mono-chromatic source, provided the separation of W_1 and W'_2 is small. However, these fringes are localized, as shown in Fig. 2.12, at the meeting point O of two rays emerging from the interferometer which are derived from a single ray incident on B_1. The position of this region of localization depends on the separation of the two rays when they leave B_2 as well as on the angle between them and can be varied by changing these parameters.

The Mach–Zehnder interferometer is a much more versatile instrument than the Michelson interferometer because each of the widely separated beam paths is traversed only once and the fringes can be localized at any

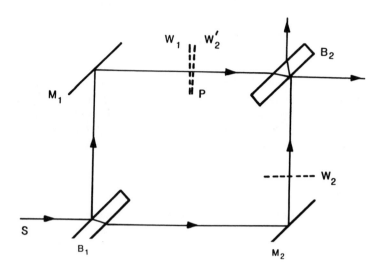

Fig. 2.11.   The Mach–Zehnder interferometer.

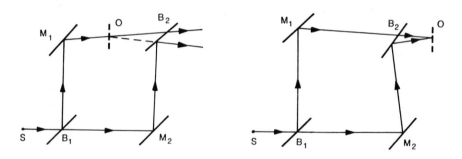

Fig. 2.12.   Localization of fringes in the Mach–Zehnder interferometer.

desired point. Because of this, it has been used extensively in studies of gas flow, combustion, plasma density and diffusion, where changes in refractive index occur which can be related to changes in pressure, temperature or the relative concentrations of different components of a mixture [see, for example, Weinberg, 1963]. However, a problem with the Mach–Zehnder interferometer is that adjusting it to give fringes of good visibility with an extended broad-band source normally involves a number of steps [Clark *et al.*, 1953; Panarella, 1973], since a displacement of a mirror results in a shift of the plane of localization as well as a change in the optical path difference; the need for this procedure can be eliminated by a modified optical arrangement [Hariharan, 1969a] which decouples the two adjustments.

2.9. The Sagnac interferometer

This type of interferometer was also first described by Michelson [Hariharan, 1975c] but is more commonly associated with Sagnac, who carried out an extensive series of experiments with it. In this interferometer, the two beams travel around the same closed circuit in opposite directions. Because of this, the interferometer is extremely stable. In addition, since the optical paths of the two beams are always very nearly equal, it is very easy to align, even with an extended broad-band source. Two forms of this interferometer are possible, as shown in Figs 2.13 (a) and (b), one with an even number of reflections in each beam and the other with an odd number of reflections.

A major difference between these is that in the form shown in Fig. 2.13 (b), the wavefronts are laterally inverted with respect to each other within the interferometer. As a result, the two beams within the interferometer can be physically separated by a lateral displacement of the incident beam. On the other hand, in the form shown in Fig. 2.13 (a), they are always superposed in the same sense.

The formation of interference fringes in an interferometer of the type shown in Fig. 2.13 (a) has been analysed in detail by Yoshihara [1968]. When B_1, M_1, M_2 are perpendicular to the plane of the figure, any two rays formed by the division of a single ray at the beam splitter will always emerge parallel to one another, though they may exhibit a lateral separation. This separation can be varied by translating any one of the elements B_1, M_1, M_2 or by rotating any one of them about an axis perpendicular to the plane of the figure. The optical path difference between two such rays at a plane normal to them is zero for a particular angle of incidence, but will increase with the deviation from this angle and the lateral separation of the rays. Consequently, when the interferometer is illuminated with a diffuse source, an observer sees a system of straight, vertical fringes localized at infinity, their spacing being inversely proportional to the lateral separation between the emerging rays.

If, however, the beam splitter is tilted about an axis in the plane of the figure, horizontal fringes can be obtained. In this case, the two emerging beams make an angle with each other, and the fringes are localized at a finite distance.

An interferometer of the form shown in Fig. 2.13 (b), with an odd number of reflections in each beam, is insensitive to displacements of the mirrors or the beam splitter. However, a rotation of any of these results in the introduction of a tilt between the two rays emerging from the interferometer derived from a single ray incident on the beam splitter. Since the lateral separation of such a pair of rays falls to zero at the mirror M_2, the fringes seen with an extended source are localized at the surface of this mirror.

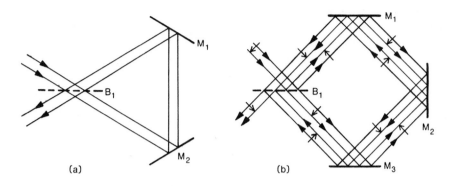

Fig. 2.13. Two forms of the Sagnac interferometer.

2.10. Interference with white light

With a point source of white light, a fringe system is produced for each wavelength, and the intensities of these fringe systems add at any point in the plane of observation. Since the path difference at the centre of the plane of observation in an interferometer such as that shown in Fig. 2.1 is zero for all wavelengths, all the fringe systems formed with different wavelengths have a maximum at this point. This results in a white central fringe. However, because the spacing of the fringes varies with the wavelength, they rapidly get out of step as the point of observation moves away from the centre of the pattern, resulting in a sequence of interference colours. These colours become less and less saturated as the path difference increases.

For interference in a thin film of air enclosed between two glass surfaces we also have to take into account the additional phase shift of π produced by reflection at one surface. As a result, the film appears black when its thickness is very small compared to the wavelength. As the thickness of the film increases, a sequence of interference colours is observed, complementary to those observed with the system shown in Fig. 2.1. Detailed calculations of the colour changes in both cases have been made by Kubota [1950, 1961]. These calculations can also be applied to interferometers using amplitude division; however, in this case, we have to take into account the fact that the phase shifts for the two beams on reflection at the beam splitter may not be the same, since, in general, these reflections do not take place under the same conditions.

2.11. Channelled spectra

With a white light source, the visibility of the fringes decreases rapidly as the optical path difference is increased. Beyond a certain point, no interference colours are seen. However, this does not mean that interference is not taking place. To illustrate this point, consider an air film of thickness d illuminated at near-normal incidence by a point source of white light. The order of interference in this film for reflected light, for any wavelength λ, is

$$N = (2d/\lambda) + (1/2) \qquad (2.57)$$

Those wavelengths which satisfy the condition

$$(2d/\lambda) = m \qquad (2.58)$$

where m is an integer, correspond to interference minima and are therefore missing in the light reflected from the film.

If, then, as shown in Fig. 2.14, an image of a small region of the film is projected on the slit of a spectroscope, the spectrum will be crossed by dark bands corresponding to the wavelengths defined by Eq. (2.58). For the wavelengths λ_1 and λ_2 corresponding to any two dark bands, we have

$$2d/\lambda_1 = m_1 \qquad (2.59)$$

and

$$2d/\lambda_2 = m_2 \qquad (2.60)$$

so that

$$d = \lambda_1 \lambda_2 (m_1 - m_2)/2 (\lambda_2 - \lambda_1) \qquad (2.61)$$

2.12. Achromatic fringes

So far, we have assumed that the optical path difference is independent of wavelength and that, consequently, the order of interference at any point is inversely proportional to the wavelength. However, it is possible to have a situation in which the optical path difference varies with wavelength in such a manner that the order of interference is very nearly independent of it. Under these conditions achromatic fringes are obtained.

Achromatic fringes can be produced if, as shown in Fig. 2.15, the secondary sources producing the interference pattern are the spectra formed by diffraction at a grating. In this case, the separation of the two images of the source S formed at S' and S'' by light of wavelength λ is

$$2a_\lambda \approx f\lambda/\Lambda \qquad (2.62)$$

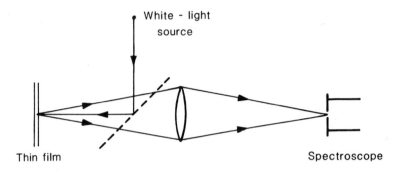

Fig. 2.14. Experimental arrangement for viewing the channelled spectrum formed by interference in a thin film.

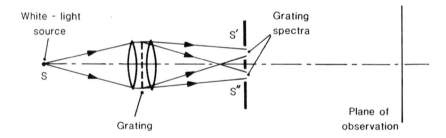

Fig. 2.15. Experimental arrangement for producing achromatic fringes.

where Λ is the spacing of the lines in the grating. Accordingly, the spacing
of the fringes in the plane of observation is

$$\Delta x = D\Lambda/2f \qquad (2.63)$$

which is independent of the wavelength. Very bright achromatic fringes can
be produced in this manner by introducing a thick tilted plate in a triangular-
path (Sagnac) interferometer [Hariharan and Singh, 1959 b].

Another method of producing achromatic fringes, or 'stationary phase'
fringes, is to introduce a dispersing medium in one of the paths in an
interferometer. As shown in Fig. 2.16, consider a Michelson interferometer
with an additional glass plate having a thickness d and a refractive index n
in one arm. The net path difference between the beams at the centre of the
field is then

$$p = p_1 - (p_2 + nd) \qquad (2.64)$$

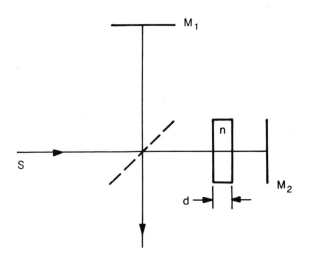

Fig. 2.16. Production of achromatic fringes in the Michelson interferometer.

where p_1 and p_2 are the paths in air in the two arms, and the interference order is

$$N = (p/\lambda)$$
$$= (1/\lambda)\,[p_1 - (p_2 + nd)] \qquad (2.65)$$

Achromatic fringes are obtained when $(dN/d\lambda) = 0$, that is to say, when

$$p_1 - \{p_2 + d\,[n - \lambda(dn/d\lambda)]\} = 0 \qquad (2.66)$$

A comparison of Eqs (2.64) and (2.66) shows that achromatic fringes can be obtained only with a finite optical path difference between the beams. With an extended source they will be circular fringes of equal inclination localized at infinity. The setting for achromatic fringes is best identified by viewing the channelled spectrum formed with a white light source. When the phase is stationary at some wavelength, the fringes at that point of the spectrum run parallel to the dispersion, or, if M_1 and M_2 are parallel, are spread out so that the interference order passes through a flat maximum [Hariharan and Singh, 1959a].

As will be shown later (see Section 3.3) the term $[n - \lambda(dn/d\lambda)]$ is the group refractive index for the glass plate. Accordingly, Eq. (2.66) is the condition that the difference of the group optical paths in the two arms is zero; this is also the condition that the interferometer is compensated for a range of wavelengths with a finite optical path difference between the beams [Steel, 1962].

2.13. Standing waves and interferential colour photography

When a parallel beam of monochromatic light is incident normally on a plane mirror, interference between the incident wave and the reflected wave results in a system of standing waves. The existence of these standing waves was demonstrated in 1890 by Wiener, who also showed that the electric vector was zero, corresponding to a node, at the reflecting surface.

For oblique incidence, the amplitude of the standing wave is a maximum when the incident light is polarized with its electric vector perpendicular to the plane of incidence, and does not then depend on the angle of incidence. If, however, the incident light is polarized in the plane of incidence, the standing wave disappears when the angle of incidence is 45°, since the electric vectors of the incident wave and the reflected wave are at right angles and they cannot interfere.

Interferential colour photography, demonstrated by Lippman in 1891, is based on recording these standing waves. If a high-resolution photographic plate is placed in contact with a mirror and illuminated normally with monochromatic light of wavelength λ_r, a series of layers of reduced silver are formed in the photographic emulsion, running parallel to its surface. These layers correspond to the antinodal planes and are separated by a distance

$$d = \lambda_r/2n \tag{2.67}$$

where n is the refractive index of the emulsion. When the processed photographic plate is illuminated with light having a wavelength λ the waves reflected from successive layers have very nearly the same amplitude and a constant optical path difference. The total reflected amplitude can then be written as

$$\begin{aligned}
u &= 1 + \exp(-i\phi) + \exp(-i2\phi) + \ldots \\
&\quad + \exp[-i(m-1)\phi] \\
&= [1 - \exp(-im\phi)]/[1 - \exp(-i\phi)]
\end{aligned} \tag{2.68}$$

where $\phi = (2\pi/\lambda)d$, and m is the number of reflecting planes. The reflected intensity is, therefore,

$$\begin{aligned}
I &= [\sin^2(m\phi/2)]/[\sin^2(\phi/2)] \\
&= [\sin^2(m\pi\lambda_r/\lambda)]/[\sin^2(\pi\lambda_r/\lambda)]
\end{aligned} \tag{2.69}$$

The reflected intensity is a maximum when $\lambda = \lambda_r$, but drops to zero when

$$\lambda - \lambda_r = \lambda_r/m \tag{2.70}$$

If m, the number of layers formed in the emulsion, is large, the plate reflects only a narrow spectral band. Each point on the plate will therefore reproduce the colour to which it was exposed.

3
Coherence

A detailed study of the effects due to the finite size and spectral bandwidth of a source requires the more powerful tools provided by coherence theory. This is essentially a statistical description of the properties of the radiation field in terms of the correlation between the vibrations at different points in the field. Comprehensive treatments of coherence theory are to be found in books by Beran and Parrent [1964], Born and Wolf [1980] and Peřina [1972], as well as in reviews by Mandel and Wolf [1965] and by Troup and Turner [1974].

3.1. Quasi-monochromatic light

As mentioned in Chapter 2, light from a thermal source is not strictly monochromatic. Accordingly, to obtain the electric field due to such a source emitting over a range of frequencies, we have to sum the fields due to the individual monochromatic components; from Eqs (2.1) and (2.9) the net electric field is given by the integral

$$V^{(r)}(t) = \int_0^\infty a(\nu) \cos [2\pi\nu t - \phi(\nu)] d\nu \qquad (3.1)$$

where $a(\nu)$ and $\phi(\nu)$ are the amplitude and phase respectively of a monochromatic component of frequency ν. Since $V^{(r)}(t)$ is a real function, our first step is to develop a complex representation which is a generalization of that used earlier for monochromatic light. For this we make use of the associated function (also real)

37

$$V^{(i)}(t) = \int_0^\infty a(\nu) \sin [2\pi\nu t - \phi(\nu)]d\nu \qquad (3.2)$$

We can then define a complex function

$$\begin{aligned} V(t) &= V^{(r)}(t) + iV^{(i)}(t) \\ &= \int_0^\infty a(\nu) \exp\{i[2\pi\nu t - \phi(\nu)]\}d\nu \end{aligned} \qquad (3.3)$$

This complex function is called the analytic signal associated with the real function $V^{(r)}(t)$; its properties have been discussed in some detail by Born and Wolf [1980]. In particular, it can be shown that if $v(\nu)$ is the Fourier transform of $V^{(r)}(t)$, so that

$$V^{(r)}(t) = \int_{-\infty}^\infty v(\nu) \exp (-i2\pi\nu t)d\nu \qquad (3.4)$$

we have

$$V(t) = 2 \int_0^\infty v(\nu) \exp (-i2\pi\nu t)d\nu \qquad (3.5)$$

It can also be shown that

$$<V(t) V^*(t)> = 2<[V^{(r)}(t)]^2> \qquad (3.6)$$

so that, if we ignore a factor of $(1/2)$, as we have done in Eq. (2.15), the optical intensity due to a quasi-monochromatic source is

$$I = <V(t) V^*(t)> \qquad (3.7)$$

It follows that if the operations on $V^{(r)}(t)$ are linear, it is possible to replace it by $V(t)$ and take the real part at the end of the calculation.

3.2. Waves and wave groups

Monochromatic light corresponds to an infinitely long train of waves all of which are identical. The superposition of infinite trains of waves of slightly different frequencies (as from a quasi-monochromatic source) results in the formation of wave groups.

For convenience, we take $E = a \cos (\omega t - kz)$, where $\omega = 2\pi\nu$ and $k = 2\pi/\lambda$, to represent a monochromatic wave propagating along the z axis. A quasi-monochromatic beam consisting of a number of such waves whose angular frequencies lie within a very small interval $\pm \Delta\omega$ about a mean angular frequency $\bar{\omega}$ can then be represented by the relation

$$V(z,t) = \int\limits_{\bar{\omega} - \Delta\omega}^{\bar{\omega} + \Delta\omega} a(\omega) \exp\left[-i\left(\omega t - kz\right)\right] d\omega \qquad (3.8)$$

If $\bar{k}$ is the propagation constant for an angular frequency $\bar{\omega}$, Eq. (3.8) can be rewritten as

$$V(z,t) = A(z,t) \exp\left[-i(\bar{\omega}t - \bar{k}z)\right] \qquad (3.9)$$

where

$$A(z,t) = \int\limits_{\bar{\omega} - \Delta\omega}^{\bar{\omega} + \Delta\omega} a(\omega) \exp\left\{-i\left[(\omega-\bar{\omega})t - (k-\bar{k})z\right]\right\} d\omega$$

$$\approx \int\limits_{\bar{\omega} - \Delta\omega}^{\bar{\omega} + \Delta\omega} a(\omega) \exp\left\{-i(\omega-\bar{\omega})\left[t-(dk/d\omega)z\right]\right\} d\omega \qquad (3.10)$$

provided $(dk/d\omega)$ does not vary appreciably over the range $\bar{\omega} \pm \Delta\omega$.

Equation (3.9) represents a wave with a single angular frequency $\bar{\omega}$ whose amplitude varies periodically at a much lower frequency. The modulation envelope effectively divides the carrier wave into groups whose length is inversely proportional to the modulation frequency and hence inversely proportional to the spectral bandwidth of the source.

3.3. Phase velocity and group velocity

Equation (3.9) shows that the planes of constant phase in an amplitude modulated wave are propagated with a velocity

$$v = \bar{\omega}/\bar{k} \qquad (3.11)$$

known as the phase velocity. The phase velocity is also the velocity of propagation of the carrier wave. On the other hand, Eq. (3.10) shows that a plane of constant amplitude is propagated with a velocity

$$v_g = d\omega/dk \qquad (3.12)$$

which is known as the group velocity.

To evaluate the group velocity, we make use of the fact that for a monochromatic wave $\omega = kv$; Eq. (3.12) then gives

$$\begin{aligned} v_g &= v + k(dv/dk) \\ &= v - \lambda(dv/d\lambda) \end{aligned} \qquad (3.13)$$

In a non-dispersive medium, $dv/d\lambda = 0$, and the group velocity is the same as the phase velocity. However, in a dispersive medium, in which the phase velocity depends on the frequency, the group velocity differs from the phase velocity.

We also have, corresponding to the group velocity, a group index of refraction

$$n_g = c/v_g \qquad (3.14)$$

which can be evaluated by expanding Eq. (3.14) as a Taylor series; we then have

$$
\begin{aligned}
n_g &= n - \lambda(dn/d\lambda) \\
&= n + \nu(dn/d\nu)
\end{aligned}
\qquad (3.15)
$$

Since, in most dispersive media, the phase velocity increases with the wavelength, $v_g < v$ and $n_g > n$. Values of n and n_g for some typical media are presented in Table 3.1.

Experimental measurements of the velocity of light from the time of flight give the group velocity and not the phase velocity, since a peak on the envelope, which identifies a particular wave train, travels with the group velocity. This distinction disappears only when the medium has zero dispersion, since then the group and phase velocities are the same. In the case of interference phenomena, the order of interference is derived from the phase index, even with quasi-monochromatic radiation; however, as will be shown later (see Section 3.11), the setting for maximum visibility of the fringes with dispersive media in the paths involves their group indices.

3.4. The mutual coherence function

The quantum theory shows that even for waves originating from a single point on a thermal source, the amplitude and phase exhibit rapid, irregular fluctuations. For waves originating from different points on a source of finite size, these fluctuations are completely uncorrelated. Because of this, the wave fields at any two points illuminated by such a source, or even at the same point at different instants of time will, in general, exhibit only partial correlation. To evaluate this correlation, consider the optical system shown in Fig. 3.1, which is similar to that used in Young's interference experiment. In this, a quasi-monochromatic source of finite size illuminates a screen containing two pinholes A_1 and A_2, and the light leaving these pinholes forms an interference pattern at a second screen.

TABLE 3.1

Phase and group indices of refraction for some materials (λ = 589 nm)

Material	n	n_g
Air	1.000 276	1.000 284
Water	1.333	1.351
Crown glass	1.517	1.542
Carbon disulphide	1.628	1.727

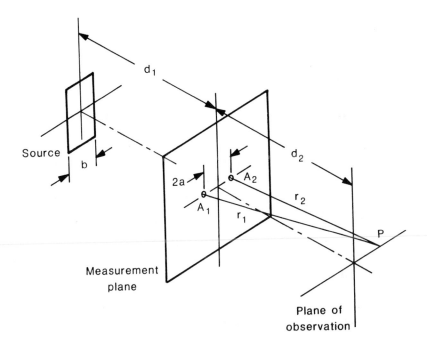

Fig. 3.1. Measurement of the coherence of the radiation field produced by a source of finite size.

Let the analytic signals corresponding to the wave-fields produced by the source at A_1 and A_2 be $V_1\,(t)$ and $V_2\,(t)$, respectively. A_1 and A_2 can then be considered as two secondary sources so that the analytic signal at a point P in the interference pattern is

$$V_P\,(t) = K_1 V_1(t-t_1) + K_2 V_2(t-t_2) \tag{3.16}$$

where $t_1 = r_1/c$ and $t_2 = r_2/c$ are the times taken for the waves from A_1 and A_2 to travel to P, and K_1 and K_2 are parameters whose magnitude is determined by geometrical factors such as the size of the apertures and the distance from the screen to P.

Now, since the interference field is stationary (or, in other words, independent of the time origin selected) Eq. (3.16) can be rewritten as

$$V_P(t) = K_1 V_1(t+\tau) + K_2 V_2(t) \tag{3.17}$$

where $\tau = t_2 - t_1$. The intensity at P is then

$$\begin{aligned}
I_P &= < V_P(t)\, V_P^*(t)> \\
&= |K_1|^2 < V_1(t+\tau)\, V_1^*(t+\tau) > + |K_2|^2 <V_2(t)\, V_2^*(t)> \\
&\quad + K_1 K_2^* < V_1(t+\tau)\, V_2^*(t) > + K_1^* K_2 < V_1^*(t+\tau)\, V_2(t)> \\
&= |K_1|^2\, I_1 + |K_2|^2\, I_2 + 2|K_1 K_2|\, \mathrm{Re}\, \Gamma_{12}(\tau) \tag{3.18}
\end{aligned}$$

where I_1 and I_2 are the intensities at A_1 and A_2, respectively, and

$$\Gamma_{12}(\tau) = < V_1(t+\tau)\, V_2^*(t)> \tag{3.19}$$

is known as the mutual coherence function of the wavefields at A_1 and A_2.

The mutual coherence function has the same dimensions as intensity. If we normalize $\Gamma_{12}(\tau)$, we obtain the dimensionless quantity

$$\gamma_{12}(\tau) = \Gamma_{12}(\tau)/(I_1 I_2)^{1/2} \tag{3.20}$$

which is called the complex degree of coherence of the wavefields at A_1 and A_2.

Now, $|K_1|^2\, I_1$ and $|K_2|^2\, I_2$ are the intensities at P due to the sources A_1 and A_2 acting separately. If we denote these intensities by $I_{P(1)}$ and $I_{P(2)}$ respectively, and make use of Eq. (3.20), Eq. (3.18) can be rewritten as

$$I_P = I_{P(1)} + I_{P(2)} + 2\,[I_{P(1)}\, I_{P(2)}]^{1/2}\, \mathrm{Re}\, \gamma_{12}(\tau) \tag{3.21}$$

This is the general equation for interference for partially coherent light.

Equation (3.21) shows that to determine the intensity at P, we must know the intensities due to A_1 and A_2 acting separately, and in addition, the real part of the complex degree of coherence. For this, we express the complex degree of coherence as the product of a modulus and a phase factor, so that

$$\gamma_{12}(\tau) = m_{12}(\tau)\, \exp\{i[\alpha_{12}(\tau) - 2\pi\bar{\nu}\tau]\} \tag{3.22}$$

where $m_{12}(\tau) = |\gamma_{12}(\tau)|$, $\alpha_{12}(\tau)$ is the phase difference between the waves incident at A_1 and A_2 and $\bar{\nu}$ is the mean frequency.

The quantity $m_{12}(\tau)$ is known as the degree of coherence. It can be shown from Eq. (3.18) and the Schwarz inequality that it satisfies the condition

$$0 \leqslant m_{12}(\tau) \leqslant 1 \qquad (3.23)$$

The wavefields at A_1 and A_2 are said to be coherent if $m_{12} = 1$, and incoherent if $m_{12} = 0$. In all other cases they are partially coherent. It should be noted that the first limiting case, $m_{12} = 1$, is attainable only for strictly monochromatic radiation and hence does not exist in reality; it can also be shown that a nonzero radiation field for which $m_{12} = 0$ for all values of τ and any pair of points in space cannot exist.

If we make use of Eq. (3.22), Eq. (3.21) becomes

$$I_P = I_{P(1)} + I_{P(2)} + 2\,[I_{P(1)}\,I_{P(2)}]^{\frac{1}{2}}\,m_{12}(\tau) \cos [\alpha_{12}(\tau) - 2\pi\bar{\nu}\tau] \quad (3.24)$$

With quasi-monochromatic light both $m_{12}(\tau)$ and $\alpha_{12}(\tau)$ are slowly varying functions of τ when compared with $\cos 2\pi\bar{\nu}\tau$ and $\sin 2\pi\bar{\nu}\tau$. As a result, when P moves across the plane of observation, the spatial variations in I_P are essentially due to the changes in $2\pi\bar{\nu}\tau$ as the value of $(r_2 - r_1)$ changes. In addition, if the openings at A_1 and A_2 are sufficiently small, we can assume that $I_{P(1)}$ and $I_{P(2)}$ are constant over an appreciable region surrounding P. The intensity distribution over this region then consists of a uniform background on which fringes are superposed corresponding to the variations of the term $\cos [\alpha_{12}(\tau) - 2\pi\bar{\nu}\tau]$.

From Eqs (2.21) and (3.24), the visibility of these fringes is

$$\mathscr{V} = 2\,[(I_{P(1)}\,I_{P(2)})^{\frac{1}{2}} \,/\, (I_{P(1)} + I_{P(2)})]\,m_{12}(\tau) \qquad (3.25)$$

which, when the two beams have the same intensity, reduces to

$$\mathscr{V} = m_{12}(\tau) \qquad (3.26)$$

In this case the visibility of the fringes gives the degree of coherence. Equation (3.24) also shows that the fringes are displaced, relative to those that would be formed if the waves from A_1 and A_2 had the same phase, by an amount proportional to the term $\alpha_{12}(\tau)$.

Another useful function which we may define at this stage is the cross-power spectrum $g_{12}(\nu)$ which is the Fourier transform of $\Gamma_{12}(\tau)$.

$$g_{12}(\nu) = \int_{-\infty}^{\infty} \Gamma_{12}(\tau) \exp (i2\pi\nu\tau) d\tau \qquad (3.27)$$

Since $\Gamma_{12}(\tau)$ is also an analytic signal, it has no components of negative frequency, and we can write the inverse transform as

$$\Gamma_{12}(\tau) = \int_{0}^{\infty} g_{12}(\nu) \exp (-i2\pi\nu\tau) d\nu \qquad (3.28)$$

For quasi-monochromatic radiation, the cross-power spectrum $g_{12}(\nu)$ has a value differing significantly from zero only over a very small frequency range on either side of the mean frequency $\bar{\nu}$. Accordingly, Eq. (3.28) can be rewritten as

$$\Gamma_{12}(\tau) = \exp\left(-i2\pi\bar{\nu}\tau\right) \int_{-\bar{\nu}}^{\infty} j_{12}(v)\exp\left(-i2\pi v\tau\right)dv$$

$$= \exp\left(-i2\pi\bar{\nu}\tau\right) J_{12}(\tau) \tag{3.29}$$

where $v = \nu - \bar{\nu}$, $j_{12}(v) = g_{12}(v)$ and $J_{12}(\tau)$, the Fourier transform of $j_{12}(v)$, is known as the mutual intensity.

Since for quasi-monochromatic light $j_{12}(v)$ differs significantly from zero only over a narrow range of values of v around $v = 0$, $J_{12}(\tau)$ contains only low-frequency components and is therefore a slowly varying function of τ.

If, therefore, the time delay τ between the interfering beams is small enough that $|(\nu-\bar{\nu})\tau| \ll 1$ over the range of frequencies for which g_{12} (ν) is not equal to zero, $|\Gamma_{12}(\tau)|$, $|\gamma_{12}(\tau)|$ and $\alpha_{12}(\tau)$ are very nearly equal to $|\Gamma_{12}(0)|$, $|\gamma_{12}(0)|$ and $\alpha_{12}(0)$ respectively. We can then set

$$J_{12} = \Gamma_{12}(0) \tag{3.30}$$

$$\mu_{12} = \gamma_{12}(0) \tag{3.31}$$

and

$$\beta_{12} = \alpha_{12}(0) \tag{3.32}$$

and rewrite Eq. (3.21) in the form

$$I_P = I_{P(1)} + I_{P(2)} + 2[I_{P(1)} I_{P(2)}]^{\frac{1}{2}} |\mu_{12}| \cos\left(\beta_{12} - 2\pi\bar{\nu}\tau\right) \tag{3.33}$$

3.5. Spatial coherence

In the limiting case just discussed when the difference in the optical paths is small enough that effects due to the finite bandwidth radiated by the source can be neglected, we are concerned essentially with the spatial coherence of the field. This can be evaluated as follows. Consider two points P_1 and P_2 in the plane of observation illuminated, as shown in Fig. 3.2, by a source S whose dimensions are small compared to its distance from P_1 and P_2. If, then, $V_{i1}(t)$ and $V_{i2}(t)$ are the analytic signals at P_1 and P_2 due to a small element dS_i located at $S(x_s, y_s, 0)$ on the source, the analytic signals at these points due to the whole source are

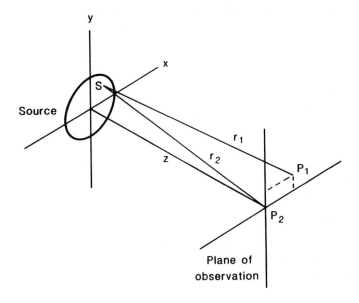

Fig. 3.2. Calculation of the coherence of the fields at two points P_1, P_2 illuminated by an extended source

$$V_1(t) = \sum_{}^{N} V_{i1}(t) \qquad (3.34)$$

and

$$V_2(t) = \sum_{}^{N} V_{i2}(t) \qquad (3.35)$$

respectively. Accordingly, the mutual intensity is

$$J_{12} = \langle V_1(t)V_2^*(t)\rangle$$

$$= \sum_{}^{N} \langle V_{i1}(t)\, V_{i2}^*(t)\rangle + \sum_{i \neq j}^{N}\sum_{}^{N} \langle V_{i1}(t)V_{j2}^*(t)\rangle \qquad (3.36)$$

However, since the fluctuations of the wave fields from different elements are statistically independent

$$\sum_{i \neq j}^{N}\sum_{}^{N} \langle V_{i1}(t)V_{j2}^*(t)\rangle = 0 \qquad (3.37)$$

so that

$$J_{12} = \sum_{}^{N} \langle V_{i1}(t)\, V_{i2}^*(t)\rangle \qquad (3.38)$$

Now, the analytic signals at P_1 and P_2 due to the element dS_i can be written as

$$V_{i1}(t) = (1/r_1) \, a_i(t - r_1/c) \exp[-i2\pi\bar{\nu}(t - r_1/c)] \qquad (3.39)$$

and

$$V_{i2}(t) = (1/r_2) \, a_i(t - r_2/c) \exp[-i2\pi\bar{\nu}(t - r_2/c)] \qquad (3.40)$$

where $a_i(t)$ is the complex amplitude of the wave field at the element dS_i at time t, and r_1 and r_2 are the distances from dS_i to P_1 and P_2. Accordingly,

$$\begin{aligned} V_{i1}(t)V_{i2}^*(t) &= (1/r_1 r_2) \, a_i(t - r_1/c) \, a_i(t - r_2/c) \\ &\times \exp[-i2\pi\bar{\nu}(r_2 - r_1)/c] \end{aligned} \qquad (3.41)$$

As shown in Section 3.1, with a quasi-monochromatic source emitting only over a range of frequencies $\bar{\nu} \pm \Delta\nu$, $a_i(t)$ varies slowly enough that if the time interval $(r_2 - r_1)/c$ is small compared to $1/\Delta\nu$, we can take $a_i(t - r_1/c) \approx a_i(t - r_2/c)$. We can therefore write

$$V_{i1}(t) \, V_{i2}^*(t) = (1/r_1 r_2) \, |a_i(t)|^2 \exp[-i2\pi\bar{\nu}(r_2 - r_1)/c] \qquad (3.42)$$

To obtain the mutual intensity due to the whole source, Eq. (3.42) is integrated over the area of the source. If $I(S)$ is the local value of the intensity at the element dS_i

$$I(S) \, dS = |a_i(t)|^2 \qquad (3.43)$$

and

$$J_{12} = \int_S (1/r_1 r_2) \, I(S) \exp[-i\bar{k}(r_2 - r_1)] \, dS \qquad (3.44)$$

where $\bar{k} = 2\pi\bar{\nu}c = 2\pi/\bar{\lambda}$.

Now, if I_1 and I_2 are the intensities at P_1 and P_2 due to the source,

$$I_1 = \int_S [I(S)/r_1^2] \, dS \qquad (3.45)$$

and

$$I_2 = \int_S [I(S)/r_2^2] \, dS \qquad (3.46)$$

Hence, from Eqs (3.44), (3.45) and (3.46), the complex degree of coherence of the fields at P_1 and P_2 is

$$\mu_{12} = (I_1 I_2)^{-\frac{1}{2}} \int_S (1/r_1 r_2)\, I(S)\, \exp[-i\bar{k}(r_2 - r_1)] dS \qquad (3.47)$$

It is apparent from the formal similarity of Eq. (3.47) to the Fresnel-Kirchhoff integral (see Appendix A.3), that the evaluation of the complex degree of coherence is equivalent to the calculation of the complex amplitude in a diffraction pattern. This result was first established by van Cittert [1934] and later obtained in a simpler way by Zernike [1938], and is now known as the van Cittert–Zernike theorem; it may be stated as follows:

> Imagine that the source is replaced by a screen with a transmittance for amplitude at any point proportional to the intensity at this point in the source, and that this screen is illuminated by a spherical wave converging to a fixed point P_2. The complex degree of coherence μ_{12} which exists between the vibrations at P_1 and P_2 is then proportional to the complex amplitude at P_1 in the diffraction pattern.

Equation (3.47) can be simplified if the dimensions of the source and the distance of P_1 from P_2 are extremely small compared to the distance of P_1 and P_2 from the source. For convenience we will assume that the point P_2 is located on the z axis at $(0.0, z)$ while $P_1(x, y, z)$ is free to move over the plane of observation. If, then, $(x_s, y_s, 0)$ are the coordinates of the point S on the source

$$r_1^2 = (x - x_s)^2 + (y - y_s)^2 + z^2 \qquad (3.48)$$

Since x_s, y_s, x, y are all very small compared to z,

$$r_1 \approx z + [(x - x_s)^2 + (y - y_s)^2]/2z \qquad (3.49)$$

Similarly, for the point P_2,

$$r_2 \approx z + (x_s^2 + y_s^2)/2z \qquad (3.50)$$

Accordingly, from Eqs (3.49) and (3.50),

$$r_2 - r_1 \approx -[(x^2 + y^2)/2z] + [(xx_s + yy_s)/z] \qquad (3.51)$$

If we define new coordinates in the source plane such that

$$\xi = x_s/z, \qquad\qquad \eta = y_s/z, \qquad (3.52)$$

and set

$$\phi_{12} = -k \, (x^2 + y^2)/2z, \tag{3.53}$$

the complex degree of coherence of the field defined by Eq. (3.47) can be written as

$$\mu_{12} = (I_1 I_2)^{1/2} \exp(i\phi_{12}) \iint_S (1/z^2) I(\xi,\eta) \exp[i\overline{k}(x\xi + y\eta)] d\xi d\eta. \tag{3.54}$$

However,

$$I_1 \approx I_2 \approx \iint_S (1/z^2) \, I(\xi,\eta) d\xi d\eta, \tag{3.55}$$

so that Eq. (3.54) becomes

$$\mu_{12} = \frac{\exp(i\phi_{12}) \iint_S I(\xi,\eta) \exp[i\overline{k}(x\xi + y\eta)] d\xi d\eta}{\iint_S I(\xi,\eta) d\xi d\eta} \tag{3.56}$$

The complex degree of coherence is therefore given in this case by the normalized Fourier transform of the intensity distribution over the source.

One case which is of interest is that of a rectangular source. Since the fringes obtained in Young's interference experiment are parallel straight lines, it is possible to use a line source instead of a point source, provided its long dimension is parallel to the fringes. Every point on the line source is incoherent with respect to all the others, but the fringes produced by the individual point sources are displaced with respect to each other only along their length and, therefore, effectively coincide. However, with a rectangular source having a width b, as shown in Fig. 3.1, the intensity distribution across the source is given by the expression

$$I(x_s) = \text{rect}(x_s/b) \tag{3.57}$$

where

$$\text{rect}(x) = \begin{cases} 1, \text{ when } |x| \le \tfrac{1}{2} \\ 0, \text{ when } |x| > \tfrac{1}{2} \end{cases} \tag{3.58}$$

Accordingly, from Eqs (3.56) and (3.26), the visibility of the fringes is

$$\mathcal{V} = |\text{sinc}(2ab/\lambda d_1)| \tag{3.59}$$

where $\text{sinc } x = (1/\pi x) \sin(\pi x)$.

It is apparent that when b is small enough that $2ab/\lambda d_1 \ll 1$, the visibility of the fringes is close to unity. As b is increased, the visibility of the fringes decreases and becomes zero when $b = \lambda d_1/2a$. Beyond this point the sinc function is negative, so that the fringes reappear, but with reversed contrast.

Another interesting case is that of a circular source of radius r. In this case, the visibility of the fringes is

$$\mathscr{V} = 2 J_1(u)/u \qquad (3.60)$$

where $u = 4\pi ra/\lambda d_1$. Experiments showing the variation of the complex degree of coherence with the effective radius of such a source have been described by Thompson and Wolf [1957] and by Hariharan and Sen [1961a].

3.6. Temporal coherence

In the other limiting case, when the source is of very small dimensions (effectively a point source) but radiates over a range of wavelengths, we are concerned with the temporal coherence of the field. In this case, the complex degree of coherence depends only on τ, the difference in the transit times from the source to P_1 and P_2, and the mutual coherence function is merely the autocorrelation function

$$\Gamma_{11} (\tau) = <V(t+\tau) V^*(t)> \qquad (3.61)$$

The degree of temporal coherence of the field is then

$$\gamma_{11} (\tau) = <V(t+\tau) V^*(t)>/<V(t) V^*(t)> \qquad (3.62)$$

From Eqs (3.25) and (3.62), it is apparent that the degree of temporal coherence can be obtained from the visibility of the interference fringes, as the optical path difference between the interfering wave fronts is varied. This leads to the concepts of the coherence time and the coherence length of the radiation.

3.7. Coherence time and coherence length

If we make use of the Wiener-Khinchin theorem, it follows from Eq. (3.61) that $S(\nu)$, the frequency spectrum of the radiation, is given, in this case, by the Fourier transform of the mutual coherence function, so that

$$S(\nu) \leftrightarrow \Gamma_{11}(\tau) \qquad (3.63)$$

while the complex degree of coherence

$$\gamma_{11}(\tau) = \mathscr{F}\{S(\nu)\} \Big/ \int_{-\infty}^{\infty} S(\nu)d\nu \qquad (3.64)$$

Consider, now, a source which radiates over a range of frequencies $\Delta\nu$ centred on a mean frequency $\bar{\nu}$. The frequency spectrum of the radiation is then given by the function

$$S(\nu) = \text{rect}\,[(\nu - \bar{\nu})/\Delta\nu] \qquad (3.65)$$

Accordingly, from Eq. (3.64), the complex degree of coherence is

$$\gamma_{11}(\tau) = \text{sinc}\,(\tau\Delta\nu) \qquad (3.66)$$

This is a damped oscillating function whose first zero occurs for a time difference $\Delta\tau$ given by the relation

$$\Delta\tau\,\Delta\nu = 1 \qquad (3.67)$$

This time interval $\Delta\tau$ is called the coherence time of the radiation; its coherence length is defined as

$$\Delta l = c\Delta\tau = c/\Delta\nu \qquad (3.68)$$

where c is the speed of light. From Eqs (2.3) and (3.68) we also have

$$\Delta l \approx (\bar{\lambda}^2/\Delta\lambda) \qquad (3.69)$$

where $\bar{\lambda}$ is the average wavelength of the source, and $\Delta\lambda$ is the range of wavelengths emitted by it.

3.8. Combined spatial and temporal effects

To evaluate the effects observed with an extended source emitting over a finite spectral band, the van Cittert–Zernike theorem is applied to the field for each spectral component and the results are integrated over the bandwidth of the radiation. We thereby obtain the integrated mutual coherence function

$$\bar{\Gamma}_{12}(\tau) = \int_{0}^{\infty} J_{12}(\nu)\,\exp\,(-i2\pi\nu\tau)d\nu \qquad (3.70)$$

where

$$J_{12}(\nu) = \int_S (1/r_1 r_2) I(S,\nu) \exp[-i\bar{k}(r_2 - r_1)] dS \qquad (3.71)$$

and $I(S, \nu)$ is the intensity distribution over the source.

3.9. Application to a two-beam interferometer

While the most important features of the interference patterns observed with simple two-beam interferometers can often be obtained quite simply from geometric considerations as outlined in Chapter 2, the application of coherence theory makes possible a more exact analysis and leads to a better understanding of some of the effects observed. Such a treatment has been developed by Steel [1965, 1967] and is outlined in this section.

We start by defining two reference planes in the interferometer. One is, as shown in Fig. 3.3, the source plane $\mathscr{S}$, while the other is $\mathscr{O}'$, the plane of observation.

On looking through the interferometer from the source side, two images of the plane of observation are seen. As shown in Fig. 3.4, these images make an angle ϵ with each other called the tilt. In addition, O_1 and O_2, the two images of the origin O, appear at different distances, z and $z + \Delta z$. The distance Δz between O_1 and O_2 along the line of sight is called the shift. Finally, the two optical paths from C, the centre of the source, to O may differ by an amount p_0 corresponding to a time difference or delay $\tau = p_0/c$. In a simple interferometer made up only of plane mirrors the shift Δz is equal to the optical path difference p_0. However, this may not be so if the paths contain focusing systems or refractive media, since the distances z and $z + \Delta z$ to O_1 and O_2 are not actual distances but correspond to the radii of curvature of the two wavefronts from O when they reach the source plane.

It is also possible, as shown in Fig. 3.5, for the images of the plane of observation to be shifted and rotated with respect to each other and even to be of different sizes. As a result, P_1, P_2, the two images of an arbitrary point in the plane of observation will exhibit a lateral separation. The vector distance $s = P_1 P_2$ is then called the shear for P. If P_1 and P_2 have position vectors u_1 and u_2 with respect to O_1 and O_2, the images of O, the origin of the plane of observation, we then have

$$s = s_0 + u_2 - u_1 \qquad (3.72)$$

where s_0 is the shear for the origin.

The complementary representation of the interferometer is shown in Fig. 3.6. On looking through the interferometer from the plane of observation,

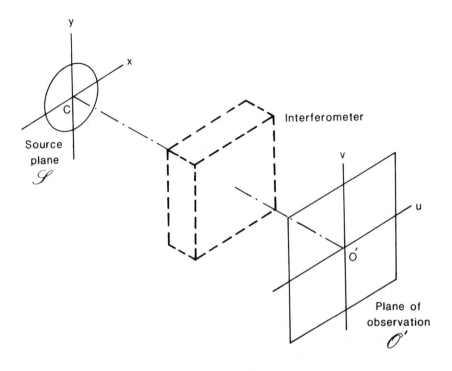

Fig. 3.3.   Equivalent optical system of an interferometer.

two images of the source are seen, whose centres C'_1 and C'_2 appear to be at distances z' and $z' + \Delta z'$. By analogy with the shift Δz defined earlier, $\Delta z'$ can be called the source shift; similarly, s', the lateral separation of S'_1 and S'_2, the images of a point S on the source, can be called the source shear for this point. We then have

$$s' = s'_0 + x'_2 - x'_1 \tag{3.73}$$

where s'_0 is the lateral separation of C'_1 and C'_2, and x'_1 and x'_2 are the position vectors of S'_1 and S'_2 with respect to C'_1, and C'_2.

To evaluate the optical path difference between the two beams from any point S on the source to any point P in the plane of observation, we first calculate the optical path difference p_0 for the central ray from C to O'; this is the difference of two summations, taken over each of the two optical paths, of the products of the thickness of individual elements and their refractive index, and can be written as

$$p_0 = \sum_2 nd - \sum_1 nd \tag{3.74}$$

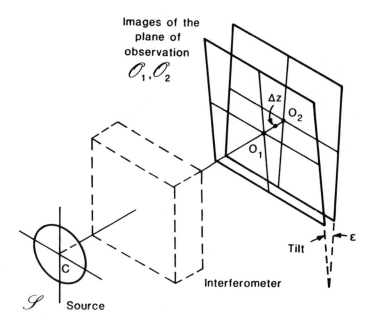

Fig. 3.4. Images of the plane of observation seen from the source side of an interferometer [Steel, 1967].

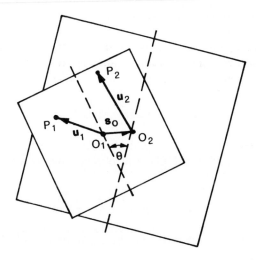

Fig. 3.5. Sheared images of the plane of observation seen from the source [Steel, 1967].

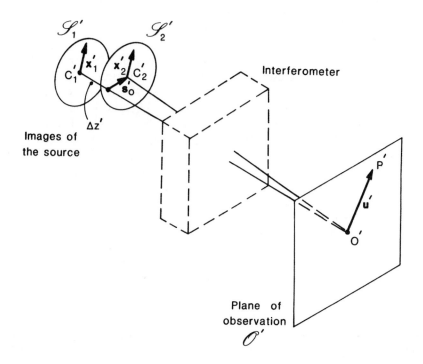

Fig. 3.6. Complementary representation of an interferometer showing the two images of the source seen from the plane of observation [Steel, 1967].

The additional optical path difference to a point P' in the plane of observation from a point S on the source, as a function of $\mathbf{x}$ the position vector of S, is then

$$\Delta p_x \approx -\mathbf{s}.\mathbf{x}/z - \Delta z |\mathbf{x}|^2 / 2z^2 \qquad (3.75)$$

Similarly, the additional optical path difference to a point S in the source plane from a point P' in the plane of observation, as a function of $\mathbf{u}'$ the position vector of P', is

$$\Delta p_u \approx -\mathbf{s}'.\mathbf{u}'/z' - \Delta z' |\mathbf{u}'|^2 / 2z'^2 \qquad (3.76)$$

Accordingly, the total path difference between the pair of points S and P' is

$$\Delta p(\mathbf{x}', \mathbf{u}') = p_0 + \Delta p_x + \Delta p_u \qquad (3.77)$$

It follows that a two-beam interferometer can be described by six parameters. Two of these are the tilt and the delay. In addition, we have, looking through the interferometer from the source side, the shift and the shear. Finally, we have, looking through the interferometer from the plane of observation, the source shift and the source shear. Diagrams corresponding to Figs 3.4 and 3.6 can be used to evaluate these parameters for any particular interferometer. They are constants for many interferometers, but in some cases they vary across the source or the plane of observation and are, therefore, functions of the position vectors x or u'. These parameters completely determine the value of $\gamma(u_1,u_2,\tau)$ the complex degree of coherence of the wave fields at any point in the plane of observation.

The intensity distribution in the plane of observation is then given by Eq. (3.21), the general equation for interference, which can now be written in the form

$$I(u') = I_1 + I_2 + 2(I_1 I_2)^{\frac{1}{2}} \, \mathrm{Re}\{\gamma(u_1,u_2,\tau)\} \tag{3.78}$$

3.10. Source-size effects

With monochromatic light, or for small path differences with quasi-monochromatic light, we can replace $\Gamma(u_1,u_2,\tau)$ in Eq. (3.78) by μ_{12} which can be evaluated from the van Cittert–Zernike theorem (see Section 3.5). We have, in this case,

$$\mu_{12} = \left\{ \int_{-\infty}^{\infty} I(x) \exp[i\phi(x)]dx \right\} / [\int_{-\infty}^{\infty} I(x)dx] \tag{3.79}$$

where $I(x)$ is the intensity distribution across the source and, from Eq. (3.75)

$$\phi(x) = \bar{k}[-s.x/z - \Delta z |x|^2 / 2z^2] \tag{3.80}$$

Equation (3.80) shows that with a point source, (for which $x = 0$) the shears and shifts have no effect on the visibility of the fringes. However, with a source of finite size, the fringes will have maximum visibility only where both the shear s and the shift Δz are very close to zero; this is the region of localization which we have found earlier by a purely geometrical analysis (see Section 2.5.2.).

As the plane of observation moves away from this region, the visibility of the fringes decreases until they vanish completely when $\mu_{12} = 0$. An interesting result, not apparent from a simple geometrical analysis, is the existence of secondary regions of localization [Hariharan, 1969b]. These arise because beyond the point at which $\mu_{12} = 0$ it is possible for its value to become finite again, though negative, and pass through a minimum. The fringes then reappear, but with a phase reversal, a dark fringe taking the place of a bright fringe.

Equations (3.79) and (3.80) also show that, to obtain fringes of good visibility with a finite shear, the source must be smaller than a certain size. From the van Cittert–Zernike theorem, this limiting size corresponds to the aperture of a lens which can barely resolve two points separated by a distance equal to the shear.

3.11. Spectral effects

With a source that also emits over a finite spectral bandwidth Eqs (3.70) and (3.78) can be used to evaluate the intensity distribution in the fringe pattern, as long as the paths traverse non-dispersive media. Conversely, the spectral energy distribution of the source can be evaluated from the Fourier transform of the self-coherence function $\Gamma_{11}(\tau)$. This was the method first used by Michelson in 1891 to study the structure of isolated spectral lines; it is also the basis of the modern technique of Fourier spectroscopy (see Chapter 11).

If, however, the beam paths include dispersive media, the delay becomes a function of the frequency and Eq. (3.78) is no longer applicable. In this case, it has been shown that, provided the group velocity can be taken to be constant over the spectral range involved, we can replace τ in Eq. (3.78) by the group delay τ_g [Pancharatnam, 1963]. The group delay is defined by analogy with Eq. (3.74) as

$$\tau_g = \sum_2 n_g d/c - \sum_1 n_g d/c \qquad (3.81)$$

where the n_g are the group refractive indices given by Eq. (3.15) for the media in the interferometer. For the fringe visibility to be a maximum, the group delay must be zero; as we have seen earlier in Section 2.13, this is the condition for achromatic fringes.

3.12. Polarization effects

To make the preceding analysis complete, we also have to evaluate the state of polarization of the beams leaving the interferometer. We note, at the outset, that if the light entering the interferometer is unpolarized or partially polarized, it can be regarded as made up of two orthogonally polarized components. Each of these can be treated separately and their intensities can be added finally.

If, then, the state of polarization of one of these orthogonal components is specified by the Jones vector A_0 (see Appendix A5), the Jones vectors A_1, A_2 of the two beams derived from it, when they leave the interferometer, are given by the relations

$$A_1 = M_1 A_{0'} \tag{3.82}$$

and

$$A_2 = M_2 A_{0'} \tag{3.83}$$

where M_1 and M_2 are matrices representing the combined effects of the elements in each of the arms. The difference in the states of polarization of the two beams can then be specified by an angle ψ defined by the relation

$$\cos \psi = <|A_1^\dagger A_2|> \tag{3.84}$$

where $A_1^\dagger$, is the Hermitian conjugate of A_1. The intensity in the interference pattern is, accordingly,

$$I(\boldsymbol{u'}) = I_1 + I_2 + 2 (I_1 I_2)^{1/2} \operatorname{Re}\{\gamma(\boldsymbol{u}_1,\boldsymbol{u}_2,\tau)\} \cos \psi \tag{3.85}$$

where

$$I_1 = <A_1^\dagger A_1> \text{ and } I_2 = <A_2^\dagger A_2>$$

A comparison with Eq. (3.78) shows that the visibility of the fringes is reduced by the factor $\cos \psi$. The generalization of this formula to quasi-monochromatic beams has been discussed by Pancharatnam [1963].

If the polarization states of the two beams leaving the interferometer are to be the same for unpolarized light (or, in other words, for any incident polarization), we require M_1 and M_2 to be identical. Each of these matrices is the product of diagonal matrices representing reflection at the mirrors and matrices which represent the change of orientation (if any) of the plane of incidence from one mirror to the next. Even if the two arms contain similar optical elements, the order in which they are traversed is not the same. As a result M_1 and M_2 will not be identical unless the matrix products commute; this requires all the component matrices to be diagonal matrices. The interferometer is then said to be compensated for polarization.

A simple case in which the interferometer is compensated for polarization is when the normals to all the mirrors and beam splitters are in the same plane. Other possible cases have been discussed by Steel [1964b]. However, compensation may not be complete if the interferometer contains elements such as a beam splitter coated with a thin metal film, for which the phase shifts on reflection differ for incidence from the air side and the glass side [Chakraborty, 1970, 1973].

Compensation for polarization is not possible in interferometers in which elements such as cube corners are used. The resulting effects, which can be quite complex, have been studied by Leonhardt [1972, 1974, 1981]. In such cases, it is possible to obtain fringes with good visibility only by using two suitably oriented polarizers in the interferometer, one just after the source and the other just before the plane of observation.

4

Multiple-beam interference

In Chapter 2, when we studied interference phenomena in plates and thin films, we neglected the contributions of multiply reflected beams. This simplification is justifiable only when the reflectance of the surfaces is low. We shall now study the fringe patterns produced when the reflectance of the surfaces is high and the multiply reflected beams must be taken into account.

4.1. Fringes in a plane-parallel plate

Consider a plane-parallel plate of refractive index n on which, as shown in Fig. 4.1, a plane wave of monochromatic light of unit amplitude is incident. As a result of multiple reflections within the plate, the incident ray SA_1, which corresponds to this wave, gives rise to a series of transmitted rays whose intensities fall off progressively. Since these rays are parallel to one another, the interference pattern produced with an extended source will be localized at infinity.

The optical path difference between successive transmitted waves can be evaluated from Eq. (2.31): the corresponding phase difference is

$$\phi = (4\pi/\lambda) \, nd \cos \theta_2 \qquad (4.1)$$

If, then, the complex reflection and transmission coefficients for amplitude at the two surfaces are r_1, t_1 and r_2, t_2, respectively, for a wave incident on

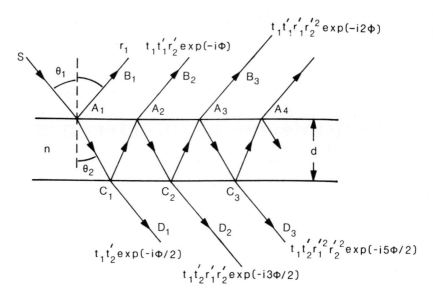

Fig. 4.1. Multiple-beam interference in a plane-parallel plate.

the surfaces from the surrounding medium, and r_1', t_1', r_2', t_2' for a wave incident on then from within the plate, the complex amplitudes of the transmitted waves $C_1D_1, C_2D_2, C_3D_3 \ldots$ can be written, as shown in Fig. 4.1, as $t_1t_2' \exp(-i\phi/2)$, $t_1t_2'r_1'r_2' \exp(-i3\phi/2)$, $t_1t_2'r_1'^2r_2'^2 \exp(-i5\phi/2) \ldots$. Accordingly, the resultant complex amplitude due to all the transmitted waves is

$$A(\phi) = t_1t_2' \exp(-i\phi/2) [1 + r_1'r_2' \exp(-i\phi)$$
$$+ r_1'^2r_2'^2 \exp(-i2\phi) + \ldots]$$
$$= t_1t_2' \exp(-i\phi/2)/[1 - r_1'r_2' \exp(-i\phi)] \qquad (4.2)$$

The complex reflection coefficients r_1', r_2' can be written in the form $r_1' = |r_1'| \exp(-i\beta_1')$, $r_2' = |r_2'| \exp(-i\beta_2')$, where β_1' and β_2' are the phase shifts on reflection at the surfaces. The intensity in the interference pattern is, therefore,

$$I(\psi) = T_1T_2/[1 + R_1R_2 - 2(R_1R_2)^{1/2} \cos\psi] \qquad (4.3)$$

where $T_1 = |t_1|^2$, $T_2 = |t_2'|^2$, $R_1 = |r_1'|^2$, $R_2 = |r_2'|^2$ are the transmittances and reflectances for intensity of the two surfaces and $\psi = \phi + \beta_1' + \beta_2'$. In the special case when $R_1 = R_2 = R$, and $T_1 = T_2 = T$, Eq. (4.3) reduces to

$$I(\psi) = T^2/[1 + R^2 - 2R \cos \psi]$$
$$= T^2/[(1-R)^2 + 4R \sin^2(\psi/2)] \tag{4.4}$$

This is known as the Airy formula.

A further simplification is possible if we are dealing with reflection at interfaces between transparent media. In this case, if the media above and below the plate are the same, the sum of the phase shifts due to successive reflections at the two surfaces, $\beta_1' + \beta_2' = 0$, or 2π, and can be neglected.

The intensity in the fringe pattern is a maximum when $\sin(\psi/2) = 0$, and a minimum when $\sin(\psi/2) = 1$. The maximum intensity is

$$I_{max} = T^2/(1-R)^2 \tag{4.5}$$

while the minimum intensity is

$$I_{min} = T^2/(1+R)^2 \tag{4.6}$$

If we define a function G such that

$$G = 4R/(1-R)^2 \tag{4.7}$$

we can write Eq. (4.4) as

$$I(\psi) = I_{max}/[1 + G \sin^2(\psi/2)] \tag{4.8}$$

and Eq. (4.6) as

$$I_{min} = I_{max}/(1+G) \tag{4.9}$$

It follows from Eq. (4.5) that if $T + R = 1$, $I_{max} = 1$. This is not the case when thin metal coatings are used to obtain increased reflectance, since these always exhibit absorption, and $T + R < 1$. However, as we will see later, reflectances approaching unity can now be obtained with multilayer dielectric films which have negligible losses and for which $T + R \approx 1$. Under these conditions, the peak intensity in the fringes is always close to unity. The intensity distribution in the fringe pattern is then given by the relation

$$I(\psi) = 1/[1 + G \sin^2(\psi/2)] \tag{4.10}$$

and is shown in Fig. 4.2 as a function of the phase difference ψ for different values of the reflectance R. As R increases, the intensity of the minima decreases, and the maxima become sharper. Near normal incidence, the

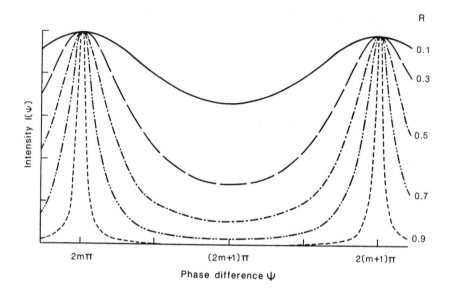

R

0.1

0.3

0.5

0.7

0.9

2mπ   (2m+1)π   2(m+1)π

Phase difference ψ

Fig. 4.2. Intensity distribution in multiple-beam fringes of equal inclination formed in trans-mitted light for different values of the reflectance.

pattern in transmitted light consists of very narrow bright circular fringes on an almost completely dark background. Such a device, consisting either of a plane-parallel plate with reflecting coatings on its two faces, or of two plates with reflecting coatings separated by an air space, is known as a Fabry-Perot interferometer or étalon.

We can define the half-width of the fringes as the interval between two points on either side of a maximum at which the intensity is equal to half its maximum value. At these points, from Eq. (4.8),

$$I_{\max}/2 = I_{\max}/[1 + G \sin^2(\psi/2)] \tag{4.11}$$

or

$$\sin(\psi/2) = G^{-\frac{1}{2}} \tag{4.12}$$

Since G is large, $\sin(\psi/2)$ is small, so that we can set $\sin(\psi/2) \approx (\psi/2)$. The half-width of the fringe then corresponds to a change in ψ of

$$W = 4G^{-\frac{1}{2}} \tag{4.13}$$

The ratio of the separation of adjacent fringes to their half-width is called their finesse. Since the separation of the fringes corresponds to a change in ψ of 2π, the finesse is

$$F = 2\pi/W = \pi G^{1/2}/2 = \pi R^{1/2}/(1-R) \qquad (4.14)$$

4.2. Fringes by reflection

In addition to the fringes produced by the transmitted beams, a fringe system is produced by the reflected beams. The resultant complex amplitude due to all the reflected waves is, from Fig. 4.1,

$$\begin{aligned}
A(\phi) &= r_1 + t_1 t_1' r_2' \exp(-i\phi)[1 + r_1' r_2' \exp(-i\phi) \\
&\quad + r_1'^2 r_2'^2 \exp(-i2\phi) + \ldots] \\
&= r_1 + [t_1 t_1' r_2' \exp(-i\phi)]/[1 - r_1' r_2' \exp(-i\phi)] \qquad (4.15)
\end{aligned}$$

Now, if we consider the simplest case of reflection at interfaces between dielectric layers, $r_1 = -r_1'$. In addition, from the Stokes's relations (see Appendix A 4.2)

$$r_1^2 + t_1 t_1' = 1 \qquad (4.16)$$

Accordingly, Eq. (4.15) can be simplified and written as

$$A(\phi) = [-r_1' + r_2' \exp(-i\phi)]/[1 - r_1' r_2' \exp(-i\phi)] \qquad (4.17)$$

In this case also, if the media above and below the plate are the same, the phase shift due to two reflections within the plate must be either 0 or 2π, and can be neglected. The intensity in the fringe pattern is, therefore,

$$I(\phi) = [R_1 + R_2 - 2(R_1 R_2)^{1/2} \cos\phi]/[1 + R_1 R_2 - 2(R_1 R_2)^{1/2} \cos\phi] \qquad (4.18)$$

In the special case when $R_1 = R_2$, Eq. (4.18) reduces to

$$\begin{aligned}
I(\phi) &= 2R(1 - \cos\phi)/(1 + R^2 - 2R\cos\phi) \\
&= 4R\sin^2(\phi/2)/[(1-R)^2 + 4R\sin^2(\phi/2)] \\
&= G\sin^2(\phi/2)/[1 + G\sin^2(\phi/2)] \qquad (4.19)
\end{aligned}$$

where, as defined by Eq. (4.7), $G = 4R/(1-R)^2$.

In this case, the pattern consists of narrow dark fringes on a nearly uniform bright background and is complementary to the pattern formed by the transmitted beams.

4.3. Fringes in a thin film: fringes of equal thickness

Very narrow fringes of equal thickness are observed between two highly reflecting surfaces enclosing a thin film; their formation has been analysed by Brossel [1947].

We consider, as shown in Fig. 4.3, a wedge-shaped film whose faces make an angle ϵ with each other, illuminated by a monochromatic plane wave W_0 incident at an angle θ. The transmitted light then contains, in addition to W_0, a family of plane waves $W_1, W_2 \ldots$ formed by multiple reflections at the film. The angle between successive members of the family is 2ϵ, and their amplitude decreases in geometric progression.

For simplicity, we will assume that the surfaces are interfaces between dielectrics. The phase shifts on reflection can then be neglected, and all the members of the family will be in phase at 0.

If we now consider a point $P(x,y)$, the optical path to the wave which has undergone $2m$ reflections is

$$
\begin{aligned}
p_m &= nPM_m \\
&= n[x \cos (\theta + 2m\epsilon) + y \sin (\theta + 2m\epsilon)]
\end{aligned}
\tag{4.20}
$$

where n is the refractive index of the film. Accordingly, the path difference between the directly transmitted wave and this wave is

$$
\begin{aligned}
\Delta p_m &= p_m - p_0 \\
&= n \{x[\cos (\theta + 2m\epsilon) - \cos \theta] + y[\sin (\theta + 2m\epsilon) - \sin \theta]\}
\end{aligned}
\tag{4.21}
$$

Equation (4.21) can be simplified for normal incidence, in which case $\theta = 0$, and the fringes are localized at the film. The point of observation P is then located in the film ($x = 0$), and we have

$$
\Delta p_m = ny \sin 2m\epsilon
\tag{4.22}
$$

If, then, the thickness of the film at the point of observation $P(0,y)$ is d we have

$$
y = d/\epsilon
\tag{4.23}
$$

and Eq. (4.22) can be expanded to give

$$
\begin{aligned}
\Delta p_m &= (nd/\epsilon) \sin 2m\epsilon \\
&= (nd/\epsilon)[2m\epsilon - (2m\epsilon)^3/3 + \ldots]
\end{aligned}
\tag{4.24}
$$

Equation (4.24) shows that with a wedged film the optical path difference between successive waves is not exactly constant, because of the presence of

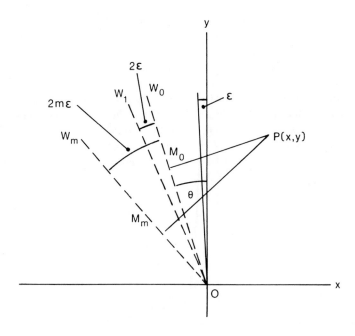

Fig. 4.3. Formation of multiple-beam fringes of equal thickness in a wedged film.

higher order terms. If these higher order terms are to be neglected, the additional path difference contributed by them must not exceed $\lambda/4$; the condition for this is

$$8m^3\epsilon^2 nd/3 < \lambda/4 \qquad (4.25)$$

This condition usually imposes severe restrictions both on the angle of the wedge and the thickness of the film.

Typically, for $\lambda = 633$ nm, if we have, say 1 fringe per mm in an air wedge, so that $\epsilon = 3.16 \times 10^{-4}$, and surfaces coated to a reflectance of 0.9, corresponding to about 50 effective beams, Eq. (4.25) gives a maximum value of d of about 5 μm.

If the thickness of the film is greater than the limit set by Eq. (4.25), the intensity distribution is no longer given by the Airy formula. Numerical calculations [Kinosita, 1953] show that the maximum intensity of the fringes drops and their width increases; in addition, they may, as shown in Fig. 4.4, suffer a displacement away from the apex of the wedge and become asymmetrical, with secondary maxima on the thicker side.

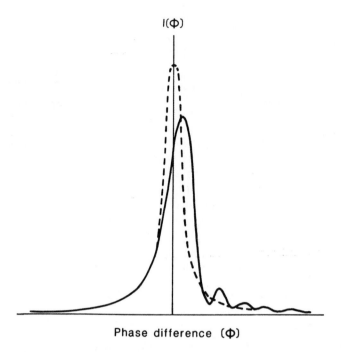

Fig. 4.4. Intensity distribution in multiple-beam fringes formed in a wedged film (solid line) compared with the Airy distribution (broken line) [Kinosita, 1953].

Fringes of equal thickness are also seen in reflected light. If there are no absorption losses, the intensity distribution in these fringes is complementary to that seen by transmission.

An important application of multiple-beam fringes of equal thickness is in the study of surface structure [Tolansky, 1955, 1961]. For this, the surface is coated with a highly reflecting layer of silver, and a small reference flat which has a semitransparent silver film on its front face is placed in contact with the surface. The reflection fringes can then be viewed by means of a microscope with a vertical illuminator.

4.4. Fringes of equal chromatic order

We have seen in Section 2.11 that when a thin film illuminated with white-light is viewed through a spectroscope, a channelled spectrum is obtained. Such fringes can also be produced by multiple-beam interference.

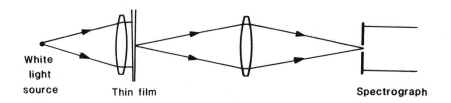

Fig. 4,5. Optical arrangement for observing fringes of equal chromatic order (FECO fringes) in a thin film

Consider the setup shown in Fig. 4.5 in which a film is illuminated at normal incidence with a collimated beam of white light and an image of the film is formed on the slit of a spectrograph. The transmitted amplitude is a maximum at wavelengths which satisfy the condition

$$(4\pi/\lambda_m)nd + \beta_1 + \beta_2 = 2m\pi \tag{4.26}$$

where β_1 and β_2 are the phase shifts on reflection at the two surfaces and m is an integer. If the reflectance of the surfaces is high, the spectrum is crossed by narrow bright fringes separated by relatively broad dark regions.

If the thickness of the film is constant, the fringes are straight lines parallel to the slit. On the assumption that the phase shifts on reflection at the surfaces do not vary with wavelength, we then have, from Eq. (4.26)

$$(1/\lambda_m) - (1/\lambda_{m+1}) = 1/2nd \tag{4.27}$$

On the other hand, if the thickness of the film varies along the section imaged on the slit, the values of λ_m along each fringe will reflect these variations. Such fringes are known as fringes of equal chromatic order [Tolansky, 1945] or FECO fringes, and have been used widely for the study of surface structure.

For this, fringes are formed between the surface under study and a flat surface. If, then, d_1 and d_2 are the thicknesses of the film at two points and the wavelengths corresponding to the fringe maximum of order m at these points are $\lambda_1(m)$ and $\lambda_2(m)$, respectively, we have, from Eq. (4.26),

$$\begin{aligned}\Delta d &= d_2 - d_1 \\ &= (m/2)\,[\lambda_2(m) - \lambda_1(m)]\end{aligned} \tag{4.28}$$

provided that the phase shifts on reflection do not change appreciably over this range of wavelengths. With a spectrograph having linear dispersion, the outline of the fringe directly corresponds to the profile of the surface. To obtain absolute values of the deviation Δd, the interference order can be evaluated from Eq. (4.27) using the values of the wavelengths corresponding to two adjacent orders, say m and $(m + 1)$, for the same point on the surface.

FECO fringes have several advantages over fringes of equal thickness for the study of surface profiles. The most important is that the wedge angle between the surfaces can be virtually eliminated, so that highly reflecting coatings can be used effectively to obtain fringes with extremely high finesse. As shown in Fig. 4.6, surface irregularities < 1 nm in height can be measured with this technique [Bennett and Bennett, 1967].

4.5. Fringes of superposition

While it is not possible to observe fringes with white light in a thick plane-parallel plate, it is possible to obtain fringes by superposing two plates of the same thickness, so that the path differences introduced in one set of beams by reflection in the first plate are matched by the path differences introduced in the other set of beams by reflection in the second plate. Such fringes were first observed by Brewster in 1817 and are called fringes of superposition.

To understand how such fringes are formed, consider two plane-parallel plates M_1, M_2 of equal thickness, which, as shown in Fig. 4.7, make a small angle ϵ with each other. Any ray incident on M_1 will give rise to a series of rays by reflection within M_1; each of these will then give rise to a series of rays by reflection within M_2. For simplicity, we select two rays B_1 and B_2 which originate from a single ray incident on M_1, B_1 being reflected at the two surfaces of M_1 and directly transmitted through M_2, while B_2 is transmitted through M_1 and reflected at the two surfaces of M_2.

Let r_1 and r_2 be the angles of refraction at the two plates. The optical path difference between these two rays is then

$$\Delta p = 2nd(\cos r_1 - \cos r_2) \qquad (4.29)$$

where n is the refractive index of the plates and d is their thickness. Since r_1 and r_2 are small, we can rewrite Eq. (4.29) as

$$\Delta p \approx 2nd[(1 - r_1^2/2) - (1 - r_2^2/2)]$$
$$\approx nd\,(r_2^2 - r_1^2) \qquad (4.30)$$

If the incident ray SA makes an angle θ with the normal to the bisector of the angle between M_1 and M_2, the angles of incidence on M_1 and M_2 are, respectively,

$$i_1 = \theta + \epsilon/2 \qquad (4.31)$$

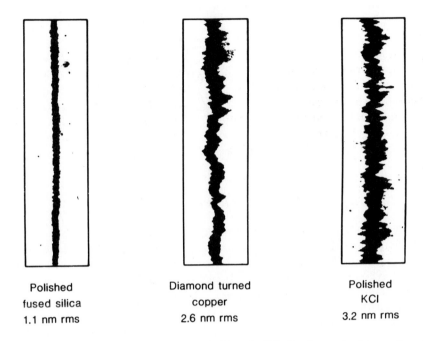

Polished	Diamond turned	Polished
fused silica	copper	KCl
1.1 nm rms	2.6 nm rms	3.2 nm rms

Fig. 4.6. Measurement of the residual irregularities of polished surfaces using fringes of equal chromatic order (courtesy J.M. Bennett, Michelson Laboratory).

and

$$i_2 = \theta - \epsilon/2 \tag{4.32}$$

However, $i_1 \approx nr_1$ and $i_2 \approx nr_2$. Accordingly, Eq. (4.30) reduces to

$$\Delta p = 2\theta\epsilon d/n \tag{4.33}$$

Equation (4.33) shows that the path difference between these two beams is proportional to θ and drops to zero when $\theta = 0$. Straight-line fringes parallel to the apex of the wedge and localized at infinity can be seen with an extended white light source, though their visibility is low because they are superposed on a uniform bright background produced by the other beams for which the optical path differences are too large for them to interfere. From Eq. (4.33), the condition for an intensity maximum is

$$2\theta\epsilon d/n = m\lambda \tag{4.34}$$

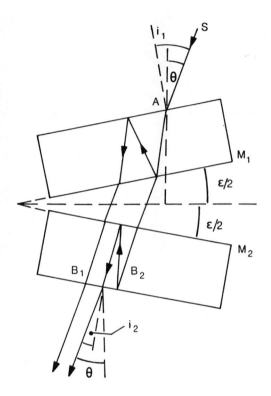

Fig. 4.7. Formation of fringes of superposition.

where m is an integer. The angular separation of the fringes is, therefore,

$$\theta_m - \theta_{m+1} = n\lambda/2\epsilon d \qquad (4.35)$$

and increases as ϵ, the angle between the plates, decreases. When the plates are parallel ($\epsilon = 0$), a uniform interference field is obtained.

A more complete analysis of this phenomenon requires us to take into account all the beams formed by multiple reflections at M_1 and M_2. In addition, we have to consider the general case in which the thicknesses of M_1 and M_2 are no longer equal.

We first consider M_1 alone and assume that a monochromatic wave of wavelength λ and unit amplitude is incident on it. If the reflectance and transmittance for amplitude for both its faces are r_1 and t_1, respectively, for a wave incident on them from the surrounding medium, and r_1' and t_1' for a wave incident from within, the complex amplitude of the transmitted wave is (see Section 4.1)

$$A_1(\lambda) = t_1 t_1' \exp(-i\phi_1/2) [1 + r_1'^2 \exp(-i\psi_1) + r_1'^4 \exp(-i2\psi_1) + \ldots]$$
$$= T_1 \exp(-i\phi_1/2) \sum_{l=0}^{\infty} R_1^l \exp(-il\psi_1) \tag{4.36}$$

where T_1 and R_1 are the transmittance and reflectance for intensity of the faces, ϕ_1 is the phase delay on transmission through the plate and ψ_1 is the phase difference between successive beams. The intensity of the transmitted wave is therefore

$$I_1(\lambda) = T_1^2 \sum_{l=0}^{\infty} \sum_{m=0}^{\infty} R_1^l R_1^m \exp[-i(l-m)\psi_1] \tag{4.37}$$

Equation (4.37) can be simplified if we set $l - m = a$; we then have

$$I_1(\lambda) = T_1^2 \sum_{l=0}^{\infty} R_1^{2l} \{1 + \sum_{a=1}^{\infty} R_1^a[\exp(ia\psi_1) + \exp(-ia\psi_1)]\}$$
$$= [T_1^2/(1-R_1^2)] (1 + 2 \sum_{a=1}^{\infty} R_1^a \cos a\psi_1) \tag{4.38}$$

In the same manner, if we consider M_2 alone, the intensity of the transmitted wave would be

$$I_2(\lambda) = [T_2^2/(1 - R_2^2)] (1 + 2 \sum_{b=1}^{\infty} R_2^b \cos b\psi_2) \tag{4.39}$$

where T_2 and R_2 are, respectively, the transmittance and reflectance for intensity of its faces, and ψ_2 is the phase difference between successive beams. Equations (4.38) and (4.39) represent the effective transmittances of M_1 and M_2. Accordingly, if we neglect the effects of beams reflected backward and forward between them, the intensity of the wave transmitted through M_1 and M_2 is the product of Eqs (4.38) and (4.39)

$$I_{1,2}(\lambda) = [T_1^2 T_2^2/(1-R_1^2)(1-R_2^2)]$$
$$\times (1 + 2 \sum_{a=1}^{\infty} R_1^a \cos a\psi_1)(1 + 2 \sum_{b=1}^{\infty} R_2^b \cos b\psi_2) \tag{4.40}$$

With a source emitting over a wide spectral bandwidth, the intensity of the transmitted wave is the sum of the intensities of the monochromatic components. If we assume that the incident light has a spectral energy distribution $I(\lambda)$, and that the reflectance and transmittance of the surfaces do not vary with wavelength, the total transmitted intensity is

$$I_{1,2} = [T_1^2 T_2^2/(1-R_1^2)(1-R_2^2)]$$
$$\times \int I(\lambda) [1 + 2 \sum_{a=1}^{\infty} R_1^a \cos a\psi_1 + 2 \sum_{b=1}^{\infty} R_2^b \cos b\psi_2$$
$$+ 2 \sum_{a=1}^{\infty} \sum_{b=1}^{\infty} R_1^a R_2^b \cos(a\psi_1 + b\psi_2)$$
$$+ 2 \sum_{a=1}^{\infty} \sum_{b=1}^{\infty} R_1^a R_2^b \cos(a\psi_1 - b\psi_2)]d\lambda \tag{4.41}$$

If the spectral bandwidth of the source is broad enough that fringes cannot be seen with either of the plates taken singly, it means that over this range of wavelengths ψ_1 and ψ_2 vary by amounts that are large compared to 2π. As a result, the terms involving $\cos a\psi_1$, $\cos b\psi_2$ and $\cos (a\psi_1 + b\psi_2)$ in Eq. (4.41) change sign many times, and their contribution to the integral can be neglected. Equation (4.41), therefore, reduces to

$$I_{1,2} = [T_1^2 T_2^2 /(1-R_1^2)(1-R_2^2)]$$

$$\times \int I (\lambda) [1 + 2 \sum_{a=1}^{\infty} \sum_{b=1}^{\infty} R_1^a R_2^b \cos (a\psi_1 - b\psi_2)] d\lambda \qquad (4.42)$$

Further consideration of Eq. (4.42) shows that the contribution of the terms involving $\cos (a\psi_1 - b\psi_2)$ is also negligible except in the particular case when ψ_1/ψ_2 is close to g/f, where g and f are two small integers. In this case, those terms involving $m(f\psi_1 - g\psi_2)$, where m is an integer, remain, and (4.42) can be written as

$$I_{1,2} = [T_1^2 T_2^2 /(1-R_1^2)(1-R_2^2)]$$

$$\times \int I(\lambda)\{1 + 2 \sum_{m=1}^{\infty} (R_1^f R_2^g)^m \cos [m(f\psi_1 - g\psi_2)]\} d\lambda \qquad (4.43)$$

The series within the braces in Eq. (4.43) is similar to that on the right-hand side of Eq. (4.38); the latter is the expression for the intensity of a mono-chromatic wave transmitted by a single plate and its sum is given by Eq. (4.10). Accordingly, we have

$$1 + 2 \sum_{m=1}^{\infty} (R_1^f R_2^g)^m \cos [m(f\psi_1 - g\psi_2)]$$

$$= [1 - (R_1^f R_2^g)^2]/\{(1-R_1^f R_2^g) + 4R_1^f R_2^g \sin^2 [(f\psi_1 - g\psi_2)/2]\} \quad (4.44)$$

and

$$I_{1,2} = \{T_1^2 T_2^2 [1 - (R_1^f R_2^g)^2]/(1-R_1^2) (1-R_2^2) (1-R_1^f R_2^g)^2\}$$

$$\times \int I (\lambda) \{1 + G \sin^2 [(f\psi_1 - g\psi_2)/2]\}^{-1} d\lambda \qquad (4.45)$$

where $G = 4R_1^f R_2^g /(1-R_1^f R_2^g)^2$

The intensity distribution defined by Eq. (4.45) is equivalent to the super-position of a number of monochromatic fringe patterns similar to that defined by Eq. (4.8). The zero-order maxima of all these patterns, which must satisfy the condition $(f\psi_1 - g\psi_2) = 0$, coincide. In the vicinity of this zero-order maximum, the intensity distribution is similar to that in a thin film whose surfaces have a reflectance $R_1^f R_2^g$. Accordingly, the contrast of the fringes decreases for higher values of f and g.

From Eq. (4.35), the angular separation of the fringes is,

$$\theta_m - \theta_{m+1} = n\lambda/2\epsilon f d_1 = n\lambda/2\epsilon g d_2 \qquad (4.46)$$

Because their separation is proportional to the wavelength, successive maxima on either side of the zero-order maximum get increasingly out of step and the visibility of the pattern falls off. With white light, the fringes on either side of the white zero-order fringe are coloured; these colours progressively become less and less distinct, and finally give way to uniform illumination. Some applications of such fringes of superposition have been discussed by Cagnet [1954].

4.6. Three-beam fringes

With two-beam interference, it is difficult to measure the position of a fringe by visual estimation to better than $1/20$ of the interfringe distance. However, visual intensity matching of two uniform interference fields, one of which contains a small phase step, makes it possible to detect a change in the optical path of $\lambda/1000$ [Kennedy, 1926; Hariharan and Sen, 1960a]. Zernike [1950] therefore proposed the following simple technique, which permits a photometric setting on a system of interference fringes.

As shown schematically in Fig. 4.8, two plane waves of equal amplitude a_0 are used as reference wavefronts; these make angles $\pm\epsilon$ with the plane of observation and intersect at a point O in this plane. The complex amplitudes due to these wavefronts at a point P in the plane of observation at a distance x from O are then

$$a_1 = a_0 \exp(-ik\epsilon x) \qquad (4.47)$$

and

$$a_2 = a_0 \exp(ik\epsilon x) \qquad (4.48)$$

and they produce an interference pattern consisting of equispaced, parallel, straight fringes.

A third wavefront which is parallel to the plane of observation is now superposed on the first two. We assume that this wave has an amplitude $2a_0$, and that its optical path differs by an amount Δz from that to the point O. The resultant complex amplitude at P can then be written as

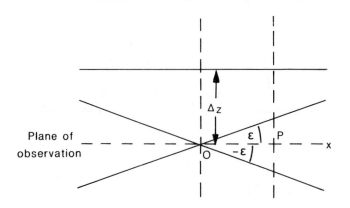

Fig. 4.8. Formation of three-beam fringes [Zernike, 1950].

$$A = a_0 \exp(-ik\epsilon x) + a_0 \exp(ik\epsilon x)$$
$$+ 2a_0 \exp(-ik\Delta z) \tag{4.49}$$

and the resultant intensity is

$$I = I_0 [3 + 2 \cos 2\psi + 4 \cos \phi \cos \psi] \tag{4.50}$$

where $I_0 = a_0^2$, $\psi = k\epsilon x$ and $\phi = k\Delta z$.

Curves of the intensity distribution in the fringes when $\phi = 2m\pi$, $2m\pi + \pi/2$ and $(2m+1)\pi$ are presented in Fig. 4.9. These show that the introduction of the third wavefront results in a modulation of the intensity of the fringes. While the minima always have zero intensity, the intensities of adjacent maxima are, in general, unequal except when ϕ is an odd multiple of $\pi/2$.

We can therefore devise a photometric setting criterion which involves adjusting the phase of the third wavefront by means of a compensator until the fringes all have the same intensity. To evaluate the precision with which this setting can be made, consider the situation when $\phi = (2m+1)\pi/2 + \Delta\phi$ where $\Delta\phi \ll \pi/2$. The intensities at adjacent maxima are then given by the relations

$$I_m = I_0 (5 - 4\Delta\phi) \tag{4.51}$$

and

$$I_{m+1} = I_0 (5 + 4\Delta\phi) \tag{4.52}$$

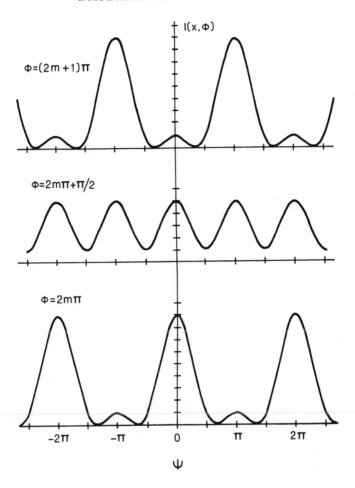

Fig. 4.9. Intensity distribution in three-beam fringes for different values of ϕ, the phase difference between the third beam and the other two beams [Hariharan and Sen, 1959].

Accordingly, the relative difference in their intensities is

$$\Delta I / I = (8/5)\Delta\phi \tag{4.53}$$

If we assume that a difference of 5% in the intensities of adjacent fringes can just be seen, the setting error is

$$\Delta\phi \approx 2\pi/200 \tag{4.54}$$

and measurements of the optical path can be made with a precision

$$\Delta p \approx \lambda/200. \qquad (4.55)$$

In the optical arrangement used by Zernike [1950], which is shown in Fig. 4.10, the three waves are produced by division of a plane wavefront at a screen containing three equidistant parallel slits. The two outer slits, whose centre lines are separated by a distance $2d$, serve to provide the reference beams, while the beam from the middle slit, which is twice as broad, is used for measurements. In this arrangement, the optical paths of all the three beams are equal at a point on the axis in the back focal plane of the lens L_2. However, at a plane located at a distance z from L_2, the optical path of the middle beam will be longer by an amount

$$\Delta p = (d^2/2)\,[(1/f) - (1/z)] \qquad (4.56)$$

There will therefore be two positions of the plane of observation that correspond to values of Δp of $\pm \lambda/4$. Any small optical path difference introduced in the middle beam can be measured from the shift in these positions. A brighter fringe system can be obtained if the source slit is replaced by a grating whose period is equal to the spacing of the fringes [Maréchal et al., 1967].

This type of interferometer is easy to set up and very stable. It has been used to calibrate phase-retardation plates as well as to measure very small changes in the optical thickness of a specimen [Vittoz, 1956]. However, because the beams are obtained by wavefront division, the amplitudes of the individual beams are not uniform over the field. As a result, the photometric setting has to be made using only the fringes on either side of the central fringe. These problems are eliminated if the three beams are produced by amplitude division using an optical system similar to that in the Jamin interferometer [Hariharan and Sen, 1959].

4.7. Double-passed fringes

In three-beam interference, the intensity distribution in a two-beam fringe pattern is modulated by the wavefront whose phase is to be measured. A similar modulation of the fringes can also be obtained by reflecting the beams emerging from an interferometer back through it [Hariharan and Sen, 1960b].

To study the formation of these fringes, consider a Michelson interferometer illuminated with collimated light, in which, as shown in Fig. 4.11,

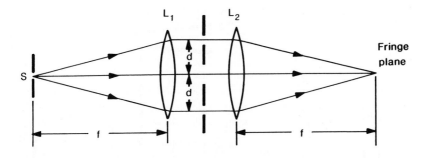

Fig. 4.10.   Zernike's three-beam interferometer.

the beams emerging from the interferometer are reflected back through it by a mirror placed at the focus of the lens L_2. The double-passed fringes are viewed, by means of an auxiliary beam splitter, at O. The beams reflected back from the interferometer after only a single pass are eliminated by the polarizer P_2, whose axis is at right angles to P_1, while the double-passed beams, which have traversed the $\lambda/4$ plate twice, are transmitted freely.

Four beams are formed in this case as a result of the double passage through the interferometer; their total optical paths may be designated as p_{11}, p_{12}, p_{21}, and p_{22} according to the paths traversed on the outward and return journeys. If, therefore, M_1 and M_2', the image of M_2 in the beam splitter, make an angle 2ϵ with each other and are separated by a distance d at the centre of the field, the optical paths of the double-passed beams at a point P at a distance x from the centre of the field can be written as

$$p_{11} = p_0 - 2d \qquad (4.57)$$

$$p_{12} = p_0 + 4x\epsilon \qquad (4.58)$$

$$p_{21} = p_0 - 4x\epsilon \qquad (4.59)$$

$$p_{22} = p_0 + 2d \qquad (4.60)$$

The resultant complex amplitude at P can then be written as

$$A = A_0 \left[\exp(i2\phi) + \exp(-i2\psi) + \exp(i2\psi) + \exp(-i2\phi) \right] \quad (4.61)$$

where $\phi = kd$ and $\psi = kx\epsilon$. The resultant intensity is, therefore,

$$I = I_0 \left[\cos 2\phi + \cos 2\psi \right]^2 \qquad (4.62)$$

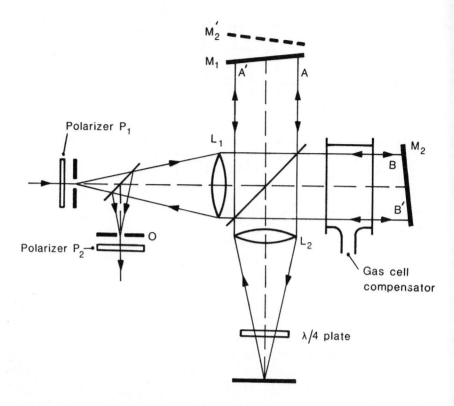

Fig. 4.11. Michelson interferometer illuminated with collimated light and double passed to obtain modulated fringes [Hariharan and Sen, 1960b].

In this case also, the intensity of adjacent fringes is equal when

$$\phi = (2m + 1)\pi/2 \qquad (4.63)$$

but the precision of measurements is

$$\Delta p = \lambda/1000 \qquad (4.64)$$

The formation of such fringes in other types of interferometers has been discussed by Hariharan and Sen [1960c, 1961b]; they have also made a detailed analysis of possible sources of error [Hariharan and Sen, 1961c].

5
Thin films

The use of thin films to produce antireflection coatings, high-efficiency mirrors and devices such as narrow-band filters now constitutes a very valuable application of optical interference [Macleod, 1969]. Interest in thin-film optics started with early observations (see Chapter 1) that the formation of a thin layer of tarnish on a glass surface results in a decrease in the amount of reflected light. We shall therefore study, in the first instance, how a thin film functions in this manner. Interesting results can be obtained with films whose refractive index varies through their depth [see, for example, Jacobsson, 1965] but we shall only consider homogeneous films.

5.1. Simple antireflection coatings

Consider a thin transparent film, whose refractive index is n_1 and whose thickness is d, on a glass substrate with a refractive index n_2, on which, as shown in Fig. 5.1, a plane wave of unit amplitude is incident. If n_0 is the refractive index of the surrounding medium, and we neglect the effects of multiple reflections, the amplitudes of the waves reflected from the two surfaces of the film are, for near normal incidence,

$$r_1 = (n_0 - n_1)/(n_0 + n_1) \tag{5.1}$$

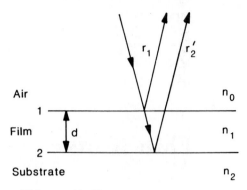

Fig. 5.1. Reflection of light at a thin film.

and

$$r_2' = (n_1 - n_2)/(n_1 + n_2) \tag{5.2}$$

If, in addition, we assume that $n_0 < n_1 < n_2$, both the waves undergo phase shifts of π on reflection, and the phase difference between them at normal incidence is

$$\phi = 4\pi n_1 d/\lambda \tag{5.3}$$

The net reflected intensity is, therefore,

$$I = r_1^2 + r_2'^2 + 2r_1 r_2' \cos \phi \tag{5.4}$$

which is a minimum when

$$n_1 d = (2m + 1)\lambda/4 \tag{5.5}$$

where m is an integer. The reflected intensity at this minimum is

$$\begin{aligned} I_{min} &= (r_1 - r_2')^2 \\ &= [2(n_0 n_2 - n_1^2)/(n_0 + n_1)(n_1 + n_2)]^2 \end{aligned} \tag{5.6}$$

which drops to zero when $n_1 = (n_0 n_2)^{1/2}$. Since the materials with the lowest refractive indices available for such coatings are $MgF_2(n = 1.38)$ and $Na_3 AlF_6$ ($n = 1.35$), this condition cannot be realized at an interface with air with any type of glass which has a refractive index less than 1.8.

For best results, the thickness of the film is adjusted so that the reflected intensity passes through a minimum at a wavelength corresponding to the peak sensitivity of the eye. Away from this wavelength, the reflectivity increases towards the red as well as the blue end of the spectrum, giving the film a residual magenta colour.

5.2. Effect of multiple reflections

To calculate the reflectance of the film accurately, we have to take into account the effect of multiple reflections in the film. The resultant complex amplitude due to all the reflected waves is, then, from Eq. (4.15),

$$A = r_1 + [t_1 t_1' r_2' \exp(-i\phi)]/[1 - r_1' r_2' \exp(-i\phi)] \tag{5.7}$$

which, since $r_1 = -r_1'$ and $r_1'^2 + t_1 t_1' = 1$ in a dielectric layer (see Appendix A.4), can be rewritten as

$$A = [r_1 + r_2' \exp(-i\phi)]/[1 + r_1 r_2' \exp(-i\phi)] \tag{5.8}$$

The reflectance of the film for intensity is therefore

$$R = AA^* = [r_1^2 + r_2'^2 + 2r_1 r_2' \cos\phi]/[1 + 2r_1 r_2' \cos\phi + r_1^2 r_2'^2] \tag{5.9}$$

The derivative of this function is

$$(dR/d\phi) = \frac{2r_1 r_2' (r_1^2 + r_2'^2 - r_1^2 r_2'^2 - 1) \sin\phi}{(1 + r_1^2 r_2'^2 + 2r_1 r_2 \cos\phi)^2} \tag{5.10}$$

which is zero, indicating a maximum or a minimum, when ϕ is zero or a multiple of π.

When $\phi = 0$ or $2m\pi$

$$R = [r_1 + r_2')/(1 + r_1 r_2')]^2$$
$$= [(n_0 - n_2)/(n_0 + n_2)]^2 \tag{5.11}$$

corresponding to the reflectance of the bare glass surface. On the other hand, when $\phi = (2m+1)\pi$

$$R = [(r_1 - r_2')/(1 - r_1 r_2')]^2$$
$$= [(n_0 n_2 - n_1^2)/(n_0 n_2 + n_1^2)]^2 \tag{5.12}$$

We can then distinguish two cases. The first is when the film has a low refractive index and

$$n_0 < n_1 < n_2 \qquad\qquad (5.13)$$

In this case, R is a maximum when $\phi = 2m\pi$ and a minimum when $\phi = (2m+1)\pi$.

The second case is when the film has a high refractive index so that

$$n_0 < n_1 > n_2 \qquad\qquad (5.14)$$

In this case, R is a minimum when $\phi = 2m\pi$ and a maximum when $\phi = (2m+1)\pi$.

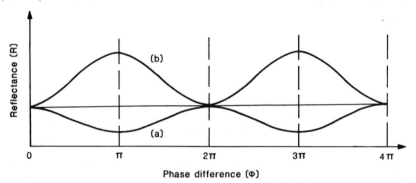

Fig. 5.2. Reflectance R plotted as a function of the phase difference ϕ for a single layer with (a) $n_0 < n_1 < n_2$, and (b) $n_0 < n_1 > n_2$.

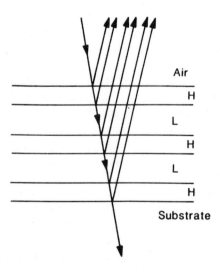

Fig. 5.3. Reflection by a multilayer stack.

The reflectance R is plotted in Fig. 5.2, as a function of the phase difference ϕ, for these two cases. In the first case, when $n_0 < n_1 < n_2$, the reflectance is represented by curve (a), which exhibits minima defined by Eq. (5.12) when $\phi = (2m + 1)\pi$. As before, the condition for the reflectance to be zero is $n_1 = (n_0 n_2)^{1/2}$. However, in the second case, when $n_0 < n_1 > n_2$, the reflectance is represented by curve (b), which exhibits maxima defined by Eq. (5.12) when $\phi = (2m + 1)\pi$. The reflectance of a glass surface can therefore be increased by depositing on it a thin transparent film with a high refractive index and an optical thickness of $\lambda/4$. Materials such as ZnS ($n = 2.35$), CeO_2 ($n = 2.2-2.4$) and TiO_2 ($n = 2.2-2.7$) have been used to produce low-loss beam splitters and have virtually replaced thin metal films for this purpose.

5.3. Multilayer stacks

Improved antireflection coatings as well as coatings with very high reflectances can be obtained with a larger number of layers. A basic type of thin film structure which is frequently used for this purpose is a stack of alternate high- (H) and low-index (L) films, all having, as shown in Fig. 5.3, an optical thickness of a quarter wavelength.

With such a stack, it is easily seen that the reflected components from all the successive boundaries emerge in phase so that their amplitudes add. As a result, it is possible to obtain a high reflectance merely by increasing the number of layers.

The reflectance of such a multilayer stack consisting of quarter-wave layers can be calculated from Eq. (5.12). A comparison of Eqs (5.12) and (5.2) shows that the reflectance of a quarter-wave layer of refractive index n_1 between media with refractive indices n_0 and n_2 is equivalent to that from the interface between two media with refractive indices of $(n_0 n_2)^{1/2}$ and n_1 respectively. Accordingly, Eq. (5.12) is applied first to the final two interfaces. These are then replaced by a single interface with the equivalent refractive indices. Equation (5.12) is then applied to this equivalent interface and the third last interface, and so on, upwards through the stack.

It is also seen from Eq. (5.11) that if a layer has an optical thickness of $m\lambda/2$ corresponding to a phase change $\phi = 2m\pi$, it has no effect on the reflectance of the surface. Accordingly, in calculating the reflectances of a multilayer system, layers whose optical thicknesses are an integral number of half wavelengths can be omitted from the calculation without affecting the result.

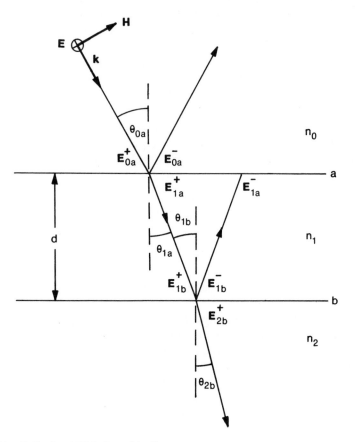

Fig. 5.4.   Reflection of light by a thin film.

5.4. Reflectance of a multilayer stack over a range of wavelengths

With a multilayer stack, it is found that the reflectance is high only over a limited range of wavelengths centred on the wavelength for which the optical thickness of the individual layers is a quarter wave. Outside this range, the reflectance drops rapidly to a low value. To evaluate the reflectance of the stack over a range of wavelengths we need a more general treatment, which is possible by solving Maxwell's equations for a stratified medium. The solution then appears as the product of two-by-two matrices, each matrix representing a single layer [Herpin, 1947; Abelès, 1950, 1967].

Consider a thin, transparent film with a refractive index n_1, bounded by two semi-infinite transparent media with refractive indices n_0 and n_2, and let a wave, linearly polarized with its electric vector perpendicular to the plane of incidence (TE−, or s −, polarization) be incident on this film as shown in Fig. 5.4. Since the electric (E) and magnetic (H) fields must be equal on both sides of the first boundary, we have

$$
\begin{aligned}
E_a &= E_{0a}^+ + E_{0a}^- \\
&= E_{1a}^+ + E_{1a}^-
\end{aligned}
\tag{5.15}
$$

where the + and − signs denote forward-going and backward-going waves, respectively, and each of the terms on the right hand side represents the resultant of all the waves of that type at this boundary. Similarly,

$$
\begin{aligned}
H_a &= (\epsilon_0/\mu_0)^{\frac{1}{2}} (E_{0a}^+ - E_{0a}^-) \, n_0 \, \cos\theta_{0a} \\
&= (\epsilon_0/\mu_0)^{\frac{1}{2}} (E_{1a}^+ - E_{1a}^-) \, n_1 \, \cos\theta_{1b}
\end{aligned}
\tag{5.16}
$$

since the electric and magnetic fields are related by the expression

$$
H = (\epsilon_0/\mu_0)^{\frac{1}{2}} \, n\hat{k} \times E
\tag{5.17}
$$

where n is the refractive index and $\hat{k}$ is the unit propagation vector.

In the same fashion, at the second boundary,

$$
\begin{aligned}
E_b &= E_{1b}^+ + E_{1b}^- \\
&= E_{2b}^+
\end{aligned}
\tag{5.18}
$$

and

$$
\begin{aligned}
H_b &= (\epsilon_0/\mu_0)^{\frac{1}{2}} (E_{1b}^+ - E_{1b}^-) \, n_1 \, \cos\theta_{1b} \\
&= (\epsilon_0/\mu_0) \, E_{2b}^+ \, n_2 \, \cos\theta_{2b}
\end{aligned}
\tag{5.19}
$$

Since the optical path traversed by the wave when it moves from one surface of the film to the other is $p_1 = n_1 d \cos\theta_{1b'}$

$$
E_{1b}^+ = E_{1a}^+ \exp(-ikp)
\tag{5.20}
$$

and

$$
E_{1b}^- = E_{1a}^- \exp(-ikp)
\tag{5.21}
$$

where $k = 2\pi/\lambda$. With these substitutions, Eqs (5.18) and (5.19) become

$$
E_b = E_{1a}^+ \exp(-ikp) + E_{1a}^- \exp(ikp)
\tag{5.22}
$$

and

$$H_b = [E_{1a}^+ \exp(-ikp) - E_{1a}^- \exp(ikp)](\epsilon_0/\mu_0)^{1/2} n_1 \cos\theta_{1b} \quad (5.23)$$

Equations (5.22) and (5.23) can be solved to obtain values for E_{1a}^+ and E_{1a}^-, and these values can be substituted in Eqs (5.15) and (5.16). We then have

$$E_a = E_b \cos(kp) + H_b i \sin(kp)/Y_1 \quad (5.24)$$

and

$$H_a = E_b Y_1 i \sin(kp) + H_b \cos(kp) \quad (5.25)$$

where

$$Y_1 = (\epsilon_0/\mu_0)^{1/2} n_1 \cos\theta_{1b} \quad (5.26)$$

is the admittance of the assembly.

Equations (5.24) and (5.25) can be written in matrix form as

$$\left[\begin{array}{c} E_a \\ H_a \end{array}\right] = \left[\begin{array}{cc} \cos(kp) & i\sin(kp)/Y_1 \\ Y_1 i\sin(kp) & \cos(kp) \end{array}\right] \left[\begin{array}{c} E_b \\ H_b \end{array}\right]$$

$$= M_I \left[\begin{array}{c} E_b \\ H_b \end{array}\right] \quad (5.27)$$

where M_I is known as the characteristic matrix relating the fields at the two boundaries.

These calculations can also be carried out for the case when the electric vector is in the plane of incidence (TM-, or p-,polarization). A similar matrix is then obtained, the only difference being that, in this case, the equivalent admittance is

$$Y_1 = (\epsilon_0/\mu_0)^{1/2} n_1/\cos\theta_{1b} \quad (5.28)$$

The use of such matrices provides a very convenient way of evaluating the reflectance of a multilayer stack. If we consider a stack consisting of j layers, the fields at the last, or ($j+1$th), boundary and the preceding (jth) boundary are related through M_j the characteristic matrix for the jth layer, so that, from Eq. (5.28)

$$\begin{bmatrix} E_j \\ H_j \end{bmatrix} = M_j \begin{bmatrix} E_{j+1} \\ H_{j+1} \end{bmatrix} \tag{5.29}$$

This calculation can be carried through, in the same manner, for all the layers in the stack; the fields at the first boundary and the last boundary are then linked by the relation

$$\begin{bmatrix} E_a \\ H_a \end{bmatrix} = M_I M_{II} \ldots \ldots M_j \begin{bmatrix} E_{j+1} \\ H_{j+1} \end{bmatrix} \tag{5.30}$$

If, instead, we represent the lumped properties of the stack by a single matrix M, which is the characteristic matrix of the entire system, then, from Eq. (5.30)

$$\begin{bmatrix} E_a \\ H_a \end{bmatrix} = M \begin{bmatrix} E_{j+1} \\ H_{j+1} \end{bmatrix} \tag{5.31}$$

where M, the characteristic matrix of the stack, is the product (taken in the proper sequence) of the characteristic matrices of the individual layers.

To derive expressions for the amplitude transmittance and reflectance of the stack in terms of the coefficients of its characteristic matrix, we first consider the case of a single layer on a substrate with a refractive index n_s and rewrite Eq. (5.27) explicitly, using the values of the fields at the two boundaries given by Eqs (5.15), (5.16), (5.18) and (5.19). We then have for the s-polarization,

$$\begin{bmatrix} E_{0a}^+ + E_{0a}^- \\ (E_{0a}^+ - E_{0a}^-)Y_0 \end{bmatrix} = M_I \begin{bmatrix} E_{2b}^+ \\ E_{2b} Y_s \end{bmatrix} \tag{5.32}$$

where

$$Y_0 = (\epsilon_0/\mu_0)^{\frac{1}{2}} n_0 \cos \theta_{0a} \tag{5.33}$$

$$Y_s \cdot = (\epsilon_0/\mu_0)^{\frac{1}{2}} n_s \cos \theta_{2b} \tag{5.34}$$

Now, $r = (E_{0a}^-/E_{0a}^+)$ and $t = (E_{2b}^+/E_{0a}^+)$. Accordingly, if

$$M_I = \begin{bmatrix} m_{11} & m_{12} \\ m_{21} & m_{22} \end{bmatrix} \tag{5.35}$$

Eq. (5.32) can be expanded and written as

$$1 + r = m_{11}t + m_{12}Y_s t \tag{5.36}$$

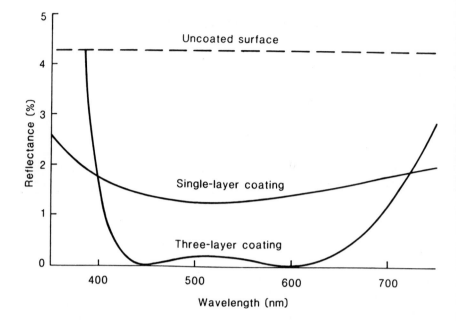

Fig. 5.5. Reflectance of a three-layer antireflection coating consisting of a high-index half-wave layer between two low-index quarter-wave layers ($n_1 = 1.38$, $n_2 = 2.15$, $n_3 = 1.62$, $n_s = 1.52$, $\lambda_{\text{design}} = 510$nm). The reflectances of the uncoated glass, as well as a single-layer antireflection coating ($n_1 = 1.38$) are also shown for comparison.

and

$$(1 - r)Y_0 = m_{21}t + m_{22}Y_s t \tag{5.37}$$

These two equations can then be solved to give

$$r = \frac{Y_0 m_{11} + Y_0 Y_s m_{12} - m_{21} - Y_s m_{22}}{Y_0 m_{11} + Y_0 Y_s m_{12} + m_{21} + Y_s m_{22}} \tag{5.38}$$

and

$$t = \frac{2Y_0}{Y_0 m_{11} + Y_0 Y_s m_{12} + m_{21} + Y_s m_{22}} \tag{5.39}$$

Equations (5.38) and (5.39) can be used to obtain the reflectance and transmittance for amplitude of a stack; all that we have to do is to replace the coefficients of the characteristic matrix for a single film in Eqs (5.38) and (5.39) by the coefficients of the characteristic matrix for the stack.

5.5. Multilayer antireflection coatings

Improved antireflection coatings are possible with multilayers. The simplest is a coating consisting of two quarter-wave layers. In this case, the condition for zero reflectance at normal incidence is

$$n_1^2 n_s / n_2^2 = n_0 \qquad (5.40)$$

This condition can be satisfied when $n_0 = 1$ and $n_s = 1.52$ with two layers having refractive indices $n_1 = 1.38$ and $n_2 = 1.70$. With three layers, a design which gives good results consists of a high-index half-wave layer inserted between two quarter-wave layers [Cox, Hass and Thelen, 1962]. Since the half-wave layer effectively disappears at the central wavelength, the refractive indices of the other two layers are the same as in the two-layer design, but, as shown in Fig. 5.5, a low reflectance is obtained over a wide range of wavelengths. Other designs for three- and four-layer antireflection coatings have also been reviewed by Musset and Thelen [1970].

A useful graphical technique in the design and production of such antireflection coatings makes use of the concept of the admittance locus [Macleod, 1972]. If we consider the deposition of a dielectric layer on a substrate and plot on an Argand diagram the locus of the admittance of this layer as its thickness increases, it is found to be a circle centred on the real axis. If we consider the case of normal incidence, one intersection of this circle with the real axis is at a point $(n_s, 0)$ where n_s is the refractive index of the substrate, while the other is at $(n_f^2 / n_s, 0)$ where n_f is the refractive index of the material of the layer. The change in the optical thickness of the layer between successive intersections is a quarter-wave, and the circle is traced out clockwise.

With a multilayer coating, the admittance locus is made up of a succession of circular arcs, each corresponding to a different layer, drawn in the order in which the layers are deposited on the substrate. At any point, the multilayer and substrate can be replaced by an equivalent admittance, which can be used to calculate the reflectance from Eq. (5.38). Obviously the reflectance of the system will be zero at the point $(n_0, 0)$, where n_0 is the refractive index of the incident medium; accordingly, for an antireflection coating, the admittance locus must terminate at this point.

5.6. Multilayer high-reflectance coatings

As mentioned in Section 5.3, a stack consisting of quarter-wave layers, alternately of high and low index, can give a very high reflectance.

Maximum reflectance is obtained when the stack consists of an odd number of layers (say $2m + 1$) with the high index layers outermost. The peak reflectance of such a stack can then be shown to be

$$R = \left[\frac{1 - (n_H/n_L)^{2m}(n_H^2/n_s)}{1 + (n_H/n_L)^{2m}(n_H^2/n_s)} \right]^2 \tag{5.41}$$

where n_H and n_L are the refractive indices of the high- and low-index layers and n_s is the refractive index of the substrate. Reflectances as high as 0.999, with an absorptance of 0.001, can be obtained with 13–15 layers.

A drawback of such multilayer stacks is that the phase shift on reflection varies with the wavelength. At the reflectance peak (at which the optical thickness of each layer is exactly a quarter wavelength), the phase shift is equal to π. However, it increases towards shorter wavelengths and decreases towards longer wavelengths. This effect has to be taken into account in interferometric measurements of wavelengths (see Chapter 7).

Another disadvantage is that the reflectance is high only over a limited range of wavelengths and falls abruptly outside this range, as shown in Fig. 5.6. For a given central wavelength, the width of this range is determined by the ratio n_H/n_L; the higher this ratio, the wider this range [Weinstein, 1954; Epstein, 1955].

5.7. Broad-band reflectors

A high reflectance can be obtained over a wider range of wavelengths by modifying the design of the stack. One method [Heavens and Liddell, 1966] is to use layers whose thickness is staggered or varies in a regular progression. This ensures that at any wavelength a sufficient number of layers have an optical thickness close enough to a quarter wavelength to give a high reflectance. Another method is to superpose two stacks which have peak reflectances at two different wavelengths, λ_1 and λ_2, chosen so that the regions of high reflectance overlap [Turner and Baumeister, 1966]. In this case, the combination exhibits a dip in reflectance at a wavelength $(\lambda_1 + \lambda_2/2)$, due to destructive interference between the waves from the two stacks. This dip can be avoided by inserting a low-index layer, with an optical thickness of a quarter wave at this wavelength, between the two stacks.

5.8. Interference filters

An interference filter is the equivalent of a Fabry-Perot interferometer (see Section 4.1) made up of thin films. Such an interferometer, when illuminated

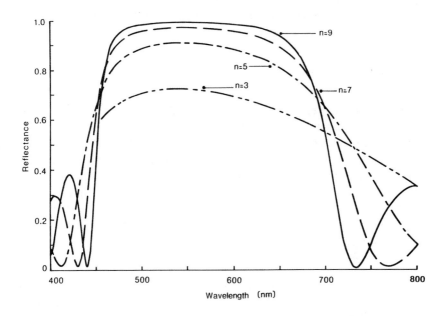

Fig. 5.6. Reflectance as a function of wavelength for quarter-wave stacks consisting of 3, 5, 7 and 9 layers ($n_H = 2.3$, $n_L = 1.38$, $n_s = 1.47$, $\lambda_{\text{design}} = 550$ nm).

with collimated light, exhibits very narrow transmission peaks at wavelengths defined by Eq. (4.10).

In its simplest form, an interference filter consists of a dielectric spacer layer sandwiched between metal reflecting layers. However, much better performance is possible if the metal layers are replaced by multilayer dielectric stacks, which can give high reflectance with negligible absorption.

Two important characteristics of an interference filter are its half-intensity bandwidth and its contrast factor. From Eq. (4.14) the half-intensity bandwidth is

$$\Delta\lambda/\lambda_0 = (1-R)/m\pi R^{\frac{1}{2}} \tag{5.42}$$

where m is the integral interference order, and R is the reflectance of the stacks. Increasing the number of layers in each stack leads to a reduction in the half-intensity bandwidth. However, Eq. (5.42) neglects the effects of the dispersion of the phase change on reflection at the multilayers; if this is taken into account, the bandwidth is reduced by a factor $(n_H - n_L)/n_H$ [Seeley, 1964].

The contrast factor of the filter is the ratio of the maximum to the minimum transmittance and, if the reflectance of the stacks does not change over the range of wavelengths considered, is, from Eqs (4.5) and (4.9),

$$C = [(1+R)/(1-R)]^2 \qquad (5.43)$$

The contrast factor also increases with the reflectance. Where necessary, a further increase in the contrast factor can be obtained with a system consisting of three highly reflecting stacks separated by two identical spacer layers; this is equivalent to two identical filters in series.

A problem with an all-dielectric filter is that since a multilayer stack has a high reflectance only over a limited range of wavelengths, it exhibits transmission sidebands at wavelengths outside this range. These can be eliminated either by auxiliary glass filters or by an induced-transmission filter, as described in the next section.

5.9. Induced-transmission filters

The transmittance of a thin metal film deposited on a transparent substrate is usually quite low over a wide range of wavelengths, because most of the incident light is reflected or absorbed. However, the characteristics of such a film can be modified by sandwiching it between two identical multilayer structures [Berning and Turner, 1957], each of which consists of a dielectric spacer layer, whose thickness is chosen to make the equivalent admittance of the metal film real, and a quarter-wave multilayer stack, which acts as an antireflection coating for a specified wavelength. At this wavelength, the overall transmission of the assembly is a maximum, limited only by absorption in the metal film. Because the stack contains a number of layers, this induced transmission is limited to a narrow wavelength region. Outside this region there is a rapid transition to enhanced absorption. No sidebands are formed on the long wavelength side, since the filter behaves as a simple metal layer in this region.

5.10. Edge filters and band-pass filters

It is possible to take advantage of the sharp edges of the region of high reflectance of a multilayer dielectric stack to obtain either a long-wave- or short-wave- pass filter. The ripples in the region of transmittance can be eliminated by modifying the design of the stack, one of the simplest modifications being the addition of an eighth-wave layer at each end of the stack. Where the range of wavelengths rejected is not broad enough, it can be extended by using either an additional absorption filter or a second stack whose rejection region overlaps that of the first. In the latter case, the two stacks must be separated by an adequate thickness of glass to avoid interference of rays reflected back and forth between the stacks.

A filter with a broad pass-band can be produced by combining two edge filters; a long-wave-pass filter and a short-wave-pass filter. To avoid the formation of transmission peaks in the rejection zones due to coupling effects, the high reflectance regions of the two edge filters should not overlap; alternatively, they can be decoupled by putting them on opposite sides of a substrate.

5.11. Dichroic beam splitters and cold mirrors

With a multilayer stack, the light which is not reflected is transmitted, the losses being negligible. It is therefore possible to use either two long-wave-pass filters or two short-wave-pass filters to produce red, green and blue colour separations from a single image.

A short-wave-pass filter which reflects all wavelengths greater than about 700 μm can be used as a heat-reflecting filter to minimize the risk of heat damage to the film in a projector. The same result can also be achieved by a cold mirror, a long-wave-pass filter which reflects only the visible wavelengths and transmits the infrared.

5.12. Polarizing beam-splitters

Away from normal incidence, the reflectance of an interface for TM waves (p-polarization) is lower than that for TE waves (s-polarization). This can be a problem in constructing a beam-splitter, and special designs are necessary if polarization effects are to be minimized [Mahlein, 1974]. However, it can be turned to advantage in the production of polarizing beam-splitters.

An early design, proposed by MacNeille and produced by Banning [1947], took advantage of the fact that the reflectance for the p-polarization vanishes at the Brewster angle. Accordingly, a high reflectance for the s-polarization can be obtained, while maintaining a transmittance close to unity for the p-polarization by the use of a multilayer stack consisting of layers whose optical thickness is a quarter-wave at this angle of incidence.

If the refractive indices of the materials used are n_H and n_L, the Brewster angle for the interface between them is defined by the condition

$$\tan \theta_H = n_L / n_H \tag{5.44}$$

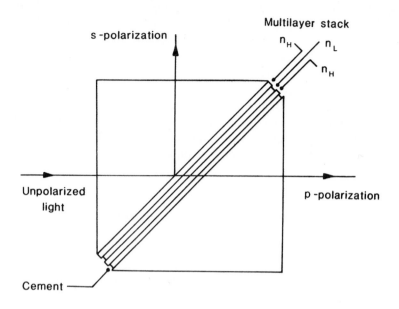

Fig. 5.7. Construction of a polarizing beam splitter.

Since a beam traversing the high index layer at this angle would be totally reflected at an interface with air, prisms are cemented on the input and exit faces of the stack as shown in Fig. 5.7. The angle of incidence in the glass (refractive index n_G) is then given by the relation

$$\sin \theta_G = (n_H / n_G) \sin \theta_H \qquad (5.45)$$

It is convenient to use right-angle prisms, in which case $\theta_G = 45°$; with $n_H = 2.35$ and $n_L = 1.35$, n_G must then be 1.66. The construction of such beam-splitters has been discussed in detail by Clapham *et al.* [1969], who have shown that a polarizing efficiency of 99.8% can be obtained.

A problem with this design is wedge fringes at the cemented interface, since the refractive index of available cements is lower than that of the prisms. A better approach [Netterfield, 1977] is to use a glass with the same refractive index as common cements (typically, 1.57) and a multilayer which gives adequate polarization selectivity under these conditions at an angle of incidence of 45°. This is possible because, with a multilayer stack at high angles of incidence, the width of the stopping zone for the *p*-polarization is much less than that for the *s*-polarization. Accordingly, the stack can be designed so that the first zero of reflectance for the *p*-polarization occurs at the wavelength at which the beam-splitter is to be used, while the reflectance for the s-polarization is still high. The arrangement then constitutes an almost ideal polarizing beam-splitter at this wavelength.

6
The laser as a light source

The high degree of spatial and temporal coherence of laser light eliminates many of the problems associated with the use of thermal sources and makes the laser a very nearly ideal light source for interferometry. Since the principles of operation of lasers have been described in detail in several books [see, for example, Bloom, 1968; Svelto, 1982], they will only be summarized briefly here.

Basically, a laser makes use of the fact that if the number of atoms in a higher (or excited) state in a medium is greater than that in the normal (or ground) state, a wave of the right frequency passing through the medium will be amplified instead of being absorbed. This situation, which is called a population inversion, can be established by a number of techniques. If the medium is a gas, a common method is to use an electric discharge through the gas; if the medium is a solid or a liquid, an effective technique is to illuminate it with an intense pulse of light from a flashlamp.

The feedback needed to convert such an amplifier into an oscillator is provided by enclosing a column of the active medium between two highly reflecting mirrors similar to those in a Fabry-Perot interferometer. When these mirrors are properly adjusted, a wave of the right frequency originating anywhere along the optical path will be returned in the correct phase along the same optical path, so that it is amplified further. An output beam can be obtained by making one of the end mirrors partially transmitting.

6.1. Gas lasers

Gas lasers are widely used for interferometry because they are cheap, easy to use and give a visible or infrared output with good spatial and temporal coherence. The principal output wavelengths available from the most common gas lasers are listed in Table 6.1. Of these, the helium–neon (He–Ne) laser, which is relatively easy to operate and, with suitable mirrors, gives a single output line at 632.8 nm is by far the most frequently used for interferometry. The helium–cadmium (He–Cd) laser is also easy to operate and gives an output in the blue–violet region of the spectrum. Ion lasers, such as the argon–ion (Ar^+) and krypton–ion (Kr^+) lasers, normally require water cooling and a wavelength-selecting prism, but are useful in applications such as holography (see Chapter 12) for which higher powers are needed. The most widely used infrared laser is the CO_2 laser, which can operate on any one of a number of lines in the 9 μm and 10.6 μm bands.

6.2. Laser modes

The two mirrors in a laser constitute a resonant cavity whose length L is much greater than its lateral dimensions. The modes of such a resonant cavity can be defined as stationary configurations of the electromagnetic field within the cavity which satisfy the boundary conditions. Obviously, since the sides of this cavity are open, only those modes involving waves travelling nearly parallel to the long axis of the resonator will have low enough losses to sustain laser oscillation. We can describe a particular configuration of this cavity as stable if an arbitrary ray moving to and fro between the two mirrors stays indefinitely within their boundaries.

The simplest form of resonant cavity is the plane-parallel or Fabry-Perot resonator. The modes of such a resonator correspond to two systems of plane waves propagating in opposite senses along the axis of the cavity and meeting the condition that the optical path for each round trip is an integral number of wavelengths. This leads to the relation

$$\nu = q(c/2L) \tag{6.1}$$

where q is an integer, for the resonant frequencies. Plane-parallel resonators were used in the earlier lasers, but they are difficult to align and only marginally stable. A better configuration consists, as shown in Fig. 6.1, of two spherical mirrors with the same radius of curvature b, separated by a distance $L = b$. Since the focal points of the two mirrors coincide, this is called a confocal resonator.

TABLE 6.1

Principal output wavelengths of gas lasers

Wavelength	Laser
422 nm	He–Cd
477 nm	Ar^+
488 nm	Ar^+
514 nm	Ar^+
633 nm	He–Ne
647 nm	Kr^+
1.15 μm	He–Ne
3.39 μm	He–Ne
10.6 μm	CO_2

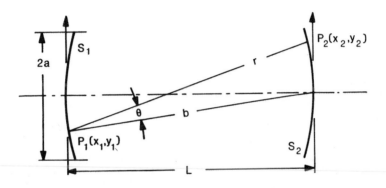

Fig. 6.1. Calculation of the modes in a confocal resonator.

6.2.1 Modes of a confocal resonator

The modes of a confocal resonator have been analysed by Boyd and Gordon [1961]. For convenience, we assume a configuration such as that shown in Fig. 6.1, with two square reflectors whose edges are parallel to the x and y axes with a length $2a$. We also assume that the field within the resonator is linearly polarized, and that its distribution over one of the reflectors (say S_1) is given by $E(x_1,y_1)$. The field at a point $P_2(x_2,y_2)$ on the other reflector S_2 can then be written, using the Fresnel–Kirchhoff integral (see Appendix A3) as

$$E(x_2,y_2) = \int_{S_1} (ik/4\pi r)(1+\cos\theta)E(x_1,y_1)\exp(-ikr)dS_1 \qquad (6.2)$$

where r is the distance between the points P_1 and P_2, θ is the angle which the line P_1P_2 makes with the normal to the mirror surface at P_1 and $k = 2\pi/\lambda$.

Since the length of the cavity is much greater than its transverse dimensions, we can set $r = L$ in the amplitude factor $(ik/4\pi r)$ and $\cos\theta = 1$, so that Eq. (6.2) reduces to

$$E(x_2,y_2) = \int_{S_1} (ik/2\pi L)E(x_1,y_1) \exp(-ikr)dS_1 \qquad (6.3)$$

Since the modes of the resonator correspond to static configurations of the field, the field distribution at the second reflector must, in its turn, generate the original field distribution at the first reflector. This is possible only if the field distributions at the two mirrors are the same, except for a factor of proportionality. We can, therefore, write

$$\sigma E(x,y) = \int_{S} (ik/2\pi L)E(x,y) \exp(-ikr)dS \qquad (6.4)$$

where σ is a complex factor of proportionality which can be written explicitly as

$$\sigma = |\sigma| \exp(i\phi) \qquad (6.5)$$

The quantity

$$\gamma = 1 - |\sigma|^2 \qquad (6.6)$$

then gives the fractional power loss on each pass due to diffraction, while ϕ is the corresponding phase delay. The condition for resonance is, obviously,

$$2\phi = 2q\pi \qquad (6.7)$$

To simplify Eq. (6.4) further, we expand r in the form of a power series. If the Fresnel number $N = a^2/L\lambda \ll L^2/a^2$, higher order terms in the expansion can be neglected, and we have

$$r = L - (1/L)(x_1x_2 + y_1y_2) \qquad (6.8)$$

If, then, we introduce the dimensionless variables

$$\xi = N^{1/2}(x/a) \qquad (6.9)$$

and

$$\eta = N^{1/2}(y/a) \qquad (6.10)$$

we can rewrite Eq. (6.4) in the form

$$\sigma\, E(\xi_2,\eta_2) = -i \int_{S_1} E(\xi_1,\eta_1) \exp\left[-i2\pi(\xi_1\xi_2 + \eta_1\eta_2)\right]d\xi_1 d\eta_1 \qquad (6.11)$$

For the case considered, namely square mirrors, the variables can be separated and Eq. (6.11) can be written as

$$\sigma_\xi E_\xi(\xi_2) = \exp\left(-i\pi/4\right) \int_{-\sqrt{N}}^{+\sqrt{N}} E_\xi(\xi_1) \exp\left(-i2\pi\xi_1\xi_2\right)d\xi_1 \qquad (6.12)$$

and

$$\sigma_\eta E_\eta(\eta_2) = \exp\left(-i\pi/4\right) \int_{-\sqrt{N}}^{+\sqrt{N}} E_\eta(\eta_1) \exp\left(-i2\pi\eta_1\eta_2\right)d\eta_1 \qquad (6.13)$$

The general solutions for these integral equations are given by the Flammer spheroidal functions [Slepian and Pollak, 1961]. However, a simple solution is possible when $N \gg 1$, so that the range of integration in Eqs (6.12) and (6.13) is effectively from $-\infty$ to $+\infty$. In this case, the right-hand sides of Eqs (6.12) and (6.13) are merely the Fourier transforms of $E_\xi(\xi_1)$ and $E_\eta(\eta_1)$, respectively. It then follows that any eigenfunction (a spatial mode) must satisfy the condition that it is its own Fourier transform.

The eigenfunctions which satisfy this condition are the product of a Gaussian function and a Hermite polynomial. They can be written in terms of the original coordinates as

$$E_{xl}(x) = H_l[x(2\pi/L\lambda)^{1/2}] \exp\left[-(\pi/L\lambda)x^2\right] \qquad (6.14)$$

and

$$E_{ym}(y) = H_m[y(2\pi/L\lambda)^{1/2}] \exp\left[-(\pi/L\lambda)y^2\right] \qquad (6.15)$$

where H_l and H_m are the Hermite polynomials of order l and m. Accordingly, the overall eigenfunction is

$$E_{lm}(x,y) = H_l H_m \exp\left[-(\pi/L\lambda)(x^2 + y^2)\right] \qquad (6.16)$$

We also know that, for resonance, the round-trip phase shift

$$kL + \phi_{lm} = q\pi \qquad (6.17)$$

It can then be shown that the resonance frequencies are given by the relation

$$\nu = (c/2L)[q + (1+l+m)/2] \qquad (6.18)$$

Equation (6.18) shows that, for any given value of q, a number of transverse modes can exist with different spatial configurations of the electric field, but with the same frequency. Since all these modes have negligible electric and magnetic fields along the axis, they are identified by a designation TEM_{lmq} where the indices l and m correspond to the orders of the Hermite polynomials H_l and H_m, and q is defined by Eq. (6.17).

The lowest-order mode, which corresponds to $l = 0$, $m = 0$ (the TEM_{00} mode) is of particular interest since its eigenfunction is

$$E_{00}(x,y) = \exp[-\pi(x^2 + y^2)/L\lambda] \qquad (6.19)$$

The field at the mirrors has a Gaussian profile, and its amplitude drops to $1/e$ of its maximum value at a distance w from the axis, known as the spot size, where

$$w = (\lambda L/\pi)^{\frac{1}{2}} \qquad (6.20)$$

This mode has the lowest losses and does not exhibit any phase reversals in the electric field across the beam, so that the output beam exhibits complete spatial coherence. For this reason, most gas lasers are designed to operate in the TEM_{00} mode.

Once the field distribution over the mirrors is known, the field distribution at any other plane inside or outside the cavity can be obtained by using the Fresnel–Kirchhoff integral. Thus, it can be shown that at the centre of the resonator the spot size is a minimum (the beam waist) and is given by the relation

$$w_0 = (\lambda L/2\pi)^{\frac{1}{2}} \qquad (6.21)$$

It can also be shown that at the beam waist the wavefront is plane, as expected from considerations of symmetry, while at the mirrors it has a radius of curvature of b.

6.2.2. Generalized spherical resonators

The confocal resonator has the disadvantage that the spot size is too small to make effective use of the available cross-section of the laser medium. For this reason, the most commonly used resonator configuration for a gas laser consists of either two concave mirrors with a radius of curvature which is greater than the resonator length, or a plane mirror in conjunction with a long-radius concave mirror. Such a resonator gives a larger spot size than a confocal resonator and can, at the same time, be aligned fairly easily.

The theory of such a generalized spherical resonator has been analysed by Boyd and Kogelnik [1962] and by Kogelnik and Li [1966] by reducing the problem to that of an equivalent confocal resonator. In this case, the frequencies of the modes are given by the relation

$$\nu_{l,m,q} = (c/2L)\{q + \frac{l+m+1}{\pi}\cos^{-1}[(1-\frac{L}{b_1})(1-\frac{L}{b_2})]^{\frac{1}{2}}\} \qquad (6.22)$$

where L is the distance between the mirrors and b_1 and b_2 are their radii of curvature, a concave surface being taken to have a positive radius. Obviously, stable modes can exist only when

$$0 \leqslant (1-L/b_1)(1-L/b_2) \leqslant 1 \qquad (6.23)$$

It should be noted that, in most lasers, the spot size at the mirrors is defined not by their dimensions but by a limiting aperture located somewhere else. In gas lasers, this is frequently the bore of the capillary used to confine the discharge. In most lasers, this aperture is (nominally) circular. The mode indices l and m then refer to the number of radial and azimuthal nodes.

6.2.3 Longitudinal modes

The active medium in a gas laser has a significant gain over a finite frequency range due to various line-broadening mechanisms such as collisions between atoms and Doppler broadening. The latter is the principal effect at low gas pressures, as in the He–Ne laser, while the former becomes dominant at higher gas pressures. Sustained oscillation is therefore possible at any frequency at which the gain in the active medium exceeds the losses in the resonator. Now, as we have seen in Section 6.2.2, even for the TEM_{00} mode, the cavity has a number of resonant frequencies which are given, for a confocal resonator, by the expression

$$\nu_q = (c/2L)(q+1/2) \qquad (6.24)$$

where q is an integer. Accordingly, if more than one of these resonant frequencies falls within the gain profile of the medium, as shown in Fig. 6.2(a), the laser, even if designed to operate in a single transverse mode, will oscillate in a number of longitudinal modes.

Low-power gas lasers can be built with the mirrors sealed directly on to the ends of the discharge tube. In this case, adjacent longitudinal modes are usually polarized at right angles. In higher power lasers, the tube containing the active medium is usually sealed with separate windows which are set at the Brewster angle. Losses for the TM wave (the wave polarized

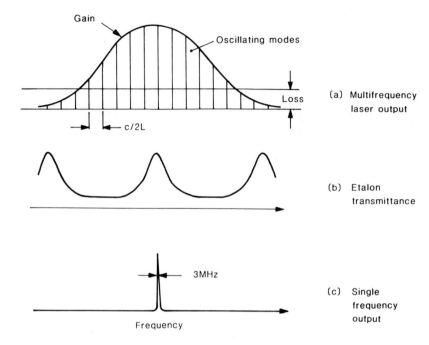

Fig. 6.2. (a) Frequency spectrum of a gas laser oscillating in a single transverse mode but in a number of longitudinal modes. (b) Transmittance of an intracavity étalon. (c) Single-frequency output obtained with an intracavity étalon.

in the plane of incidence) are then close to zero, while losses for the TE wave are high enough to prevent oscillation. As a result, all the longitudinal modes are polarized in the same plane.

The width of the individual modes depends on the mechanical stability of the cavity structure as well as on the losses and is typically about 3MHz in the visible region. With a laser oscillating in a single longitudinal mode, the coherence length could therefore be nearly 100 metres. However, the existence of more than one longitudinal mode reduces the coherence length severely. If we assume that, as a first approximation, the output spectrum of the beam can be represented by Q equally spaced delta functions, so that it can be written as

$$S(\nu) = \sum_{q}^{q+Q-1} \delta(\nu - \nu_q) \tag{6.25}$$

the degree of temporal coherence for a path difference p is

$$\gamma_{11}(p) = \left| \frac{\sin(Q\pi p/2L)}{Q \sin(\pi p/2L)} \right| \tag{6.26}$$

If $Q > 1$, this is an oscillatory function and the coherence length is defined conventionally by its first zero which occurs when

$$p = 2L/Q \qquad (6.27)$$

Fringes will, of course, be obtained once again as the optical path difference is increased beyond this point, but their contrast is reversed.

A He–Ne laser with a cavity length of 0.25 m oscillates, typically, in 2 or 3 modes separated by 600 MHz; the coherence length would therefore be between 0.16 and 0.25 m.

6.2.4 Single-frequency operation

The simplest way to obtain operation in a single longitudinal mode is to use a very short cavity. The frequency difference between the longitudinal modes is then greater than the width of the gain profile (about 1.7 GHz for a He–Ne laser) and only one longitudinal mode is sustained. The drawback of this technique is that the gain available with such a small length of the active medium is limited, so that the power output is very low. In an extreme case in which none of the modes falls within the gain profile, the laser may not even oscillate.

If single-mode operation at higher power levels is required, some method of longitudinal mode selection has to be applied to a laser with a long plasma tube. One technique described by Smith [1965] involves the use of an auxiliary resonant cavity as shown in Fig. 6.3(a). In this arrangement, the mirrors M_1 and M_2 form the main laser cavity while the combination of M_1, the beam-splitter B and the mirror M_3 can be regarded as a resonant reflector

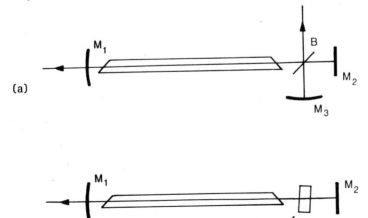

Fig. 6.3. (a) Three-mirror or Fox–Smith mode selector, and (b) intracavity étalon, for operation of a gas laser in a single longitudinal mode.

that reflects light only over a frequency band whose width is less than the separation of adjacent longitudinal modes. As a result, only one longitudinal mode has enough gain to oscillate, all other possible modes of the long cavity being rejected. This method of mode selection is very efficient and has been used successfully with He–Ne lasers as well as with Ar$^+$ lasers.

An alternative method of ensuring single frequency operation (see Fig. 6.3(b)) is to use an auxiliary étalon (usually a plane-parallel plate of fused silica) in the laser cavity [Hercher, 1969]. Oscillation is then possible only on a mode common to both cavities, as shown in Fig. 6.2 (b) and (c), for which the combined losses are low. This method works well with lasers that have a relatively high gain, such as the Ar$^+$ laser.

6.3. Comparison of laser frequencies

The coherence time for light from a laser can be long enough that it is possible to observe beats between light waves from two lasers operating on the same transition [Javan et al., 1962]. Measurements of the beat frequency and its variance over time can be used to estimate the frequency stability of a laser.

To observe such beats, the beams from the two lasers are combined at a semi-transparent mirror and detected by a photomultiplier or a silicon photodiode. The resultant electric field at the detector can then be represented by the real vibration

$$E(t) = E_1(t) + E_2(t) \tag{6.28}$$

where $E_1(t)$ and $E_2(t)$ are the real vibrations corresponding to the two superposed fields. These can be written as

$$E_1(t) = a_1 \cos(2\pi\nu_1 t + \phi_1) \tag{6.29}$$

$$E_2(t) = a_2 \cos(2\pi\nu_2 t + \phi_2) \tag{6.30}$$

where a_1 and a_2 are the amplitudes, ν_1 and ν_2 are the frequencies and ϕ_1 and ϕ_2 are the phases of the two waves.

With a square-law detector, the output current $i(t)$ is proportional to $[E(t)]^2$ so that

$$
\begin{aligned}
i(t) &= [E_1(t) + E_2(t)]^2 \\
&= (1/2)(a_1^2 + a_2^2) + (1/2)[\cos 2(2\pi\nu_1 t + \phi_1) + \cos 2(2\pi\nu_2 t + \phi_2)] \\
&\quad + a_1 a_2 \cos[2\pi(\nu_1 + \nu_2)t + \phi_1 + \phi_2] \\
&\quad + a_1 a_2 \cos[2\pi(\nu_1 - \nu_2)t + \phi_1 - \phi_2]
\end{aligned}
\tag{6.31}
$$

The second and third terms on the right-hand side of Eq. (6.31) correspond to oscillatory components at frequencies of $2v_1$, $2v_2$ and $(v_1 + v_2)$ which are too high to be followed by the detector. Accordingly Eq. (6.31) reduces to

$$i(t) = (1/2) (a_1^2 + a_2^2)$$
$$+ a_1 a_2 \cos [2\pi(v_1 - v_2)t + \phi_1 - \phi_2] \qquad (6.32)$$

The output from the detector consists of a steady current on which is superposed an oscillatory component at the beat frequency $(v_1 - v_2)$; since this is typically in the radio-frequency domain, it can be observed.

We can also interpret Eq. (6.32) as implying the formation of a set of moving interference fringes, the number of fringes passing any point on the detector in unit time being equal to the beat frequency. Obviously, for the beats to be detected, it is necessary for the dimensions of the detector to be small compared to the interfringe distance, or in other words,

$$d \ll \lambda/\sin \theta \qquad (6.33)$$

where d is the diameter of the sensitive area of the detector and θ is the angle between the wavefronts.

Beats can even be observed with a single laser if it is oscillating in more than one axial mode [Herriot, 1962]. The frequency separation of these modes and, therefore, the beat frequency is given, as we have seen earlier, by the relation

$$\Delta v = c/2L \qquad (6.34)$$

where c is the speed of light and L is the length of the laser cavity. With a typical He–Ne laser having a cavity length of 250 mm, this beat frequency is about 600 MHz and can be measured either with a spectrum analyzer using a fast photodiode as the detector, or by direct frequency counting.

6.4 Frequency stabilization

For many types of interferometric measurements, it is essential that the frequency of the source should be stable. In addition, for measurements of length (see Chapter 7) it should be precisely known. The output from a free-running gas laser, even when it is oscillating in a single longitudinal mode, normally does not meet these requirements to better than about 1 part in 10^6. The reason for this is that the output frequency is inversely proportional to the optical distance between the end mirrors which can vary due to thermal or mechanical effects. As a result, although such a laser can, in principle, produce an output with a bandwidth less than 500 Hz, the mean frequency of this output can vary over the much wider bandwidth of the gain profile. Because of this, some method of stabilizing the output frequency of a laser is necessary. The various methods of frequency stabilization available have been discussed in reviews by Wallard [1973] and by Baird and Hanes [1974] (see also the proceedings of the Third Symposium on Frequency Standards and Metrology, Aussois, France, 1981, published in *J. Phys. Colloq.* (France). 42 C-8); some of the more important methods will be described briefly here.

6.4.1 Polarization-stabilized laser

A simple method of stabilizing the output frequency of a He–Ne laser involves using a laser tube with internal mirrors, whose length is chosen so that it operates simultaneously on just two longitudinal modes. Normally, these two modes are orthogonally polarized and their directions of polar-

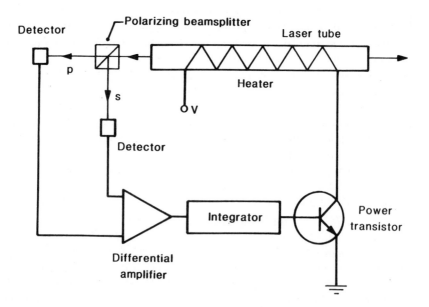

Fig. 6.4. Frequency stabilization system for a He–Ne laser using two orthogonally polarized modes [Gordon and Jacobs, 1974].

ization remain fixed. It is therefore possible to divide the beam from the rear mirror of the laser at a polarizing beam-splitter, as shown in Fig. 6.4, and compare the intensities of the two modes. The difference of the two intensities is used as the error signal in a servo amplifier that controls either the discharge current [Balhorn et al., 1972] or, as shown in Fig. 6.4, the current in a heater wound on the plasma tube [Bennett et al., 1973; Gordon and Jacobs, 1974], so that the intensities of the two modes are held equal. The frequencies of the two modes are then symmetrically positioned about the centre frequency of the gain profile. One of the modes can be selected by means of a polarizer in the main beam to obtain a single-frequency output.

Detailed measurements with such lasers show that their frequency variations can be held to less than $\pm$ 50 kHz for short periods and $\pm$ 5 MHz, or 1 part in 10^8, over quite long periods [Ciddor and Duffy, 1983]. However, care must be taken to avoid systematic offsets due to optical feedback arising from external reflections [Brown, 1981].

6.4.2 Stabilized tranverse Zeeman laser

As mentioned earlier, a low-power He–Ne laser with internal mirrors usually oscillates in two or three longitudinal modes which are linearly polarized in orthogonal directions. Morris et al. [1975] found that the application of a transverse magnetic field to such a laser results in the modes splitting. If the value of the magnetic field is properly chosen, adjacent components of two modes can be made to overlap near the line centre, while the outer components move to the wings of the gain curve and are suppressed. As a result the laser oscillates on a single axial mode composed of two orthogonally polarized waves. These two components exhibit a small frequency difference due to the magnetically induced birefringence of the gas in the laser tube, and this frequency difference depends on the position of the mode within the gain profile. This property has been exploited to stabilize the output frequency of the laser.

As shown in Fig. 6.5, the two frequencies in the back beam of the laser are mixed with a polarizer and the beat frequency is detected with a phototransistor. The output from a frequency-to-voltage converter circuit is then used to control the length of the cavity through a servo amplifier and a heating coil on the laser tube. The properties of lasers using this method of frequency stabilization have been studied in detail by Ferguson and Morris [1978] and by Umeda et al. [1980]. Brown [1981] has shown that the frequency variations are of the same order as those cited earlier for polarization stabilization but, since stabilization is effected by measurements of a beat frequency, this system is less influenced by optical feedback; however, problems can arise due to stray magnetic fields.

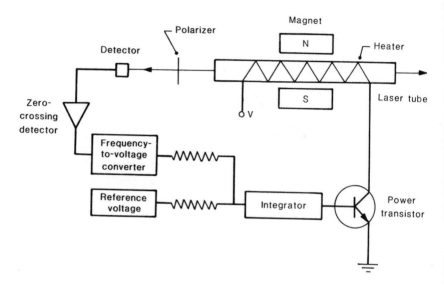

Fig. 6.5.   Frequency stabilization system for a He–Ne laser using a transverse magnetic field [Morris *et al.*, 1975].

6.4.3 Stabilization on the Lamb dip

When a laser oscillates, the number of atoms available in the upper (or excited) state drops, so that the available gain is reduced. However, this reduction is not uniformly spread over the whole gain profile. With a laser operating at a single frequency, only those atoms whose Doppler-shifted frequencies match the laser frequency within narrow limits will interact with the radiation field. As a result, there is a local dip in the gain profile; this effect is known as 'hole burning'. If the laser frequency is away from the centre of the Doppler curve, two holes are burned in the gain profile as shown in Fig. 6.6. These arise because of the interaction of waves travelling in opposite directions along the axis of the resonator with these two groups of atoms having equal but opposite Doppler shifts. However, if the laser frequency coincides with the centre of the Doppler curve, these two holes coalesce. Because of saturation of the stimulated emission, this results in a drop in the output power known as the Lamb dip. The width of the Lamb dip is about a tenth of the Doppler width for the He–Ne laser and it has been used, by modulating the laser frequency and monitoring the output with a phase-sensitive detector, to stabilize its average frequency to about 1 part in 10^8 [Rowley and Wilson, 1972].

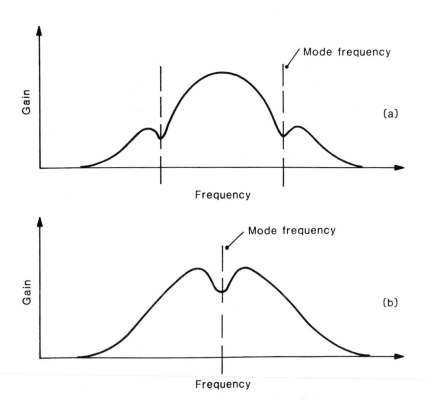

Fig. 6.6. 'Hole burning' due to saturation of stimulated emission; (a) mode frequency offset from the centre of the gain profile, and (b) mode frequency at the centre of the gain profile.

6.4.4 Stabilization by saturated absorption

Stabilization on the Lamb dip has been superseded by a similar technique based on saturated absorption. When a cell containing gas molecules with an absorption line at a suitable frequency is placed within a laser cavity, two groups of molecules with equal but opposite velocity components will normally interact with the two counter-propagating waves. However, when the laser frequency coincides with the unshifted absorption frequency, the same group of molecules will interact with the two counter-propagating waves and there is a dip in the observed absorption due to saturation of the absorption. The width of this dip is limited only by the lifetimes of the energy states involved and the interaction time, and is typically less than 10^{-3} of the Doppler width.

Frequency stabilization by this technique was first proposed by Lee and Skolnick [1967], and later achieved by Hall [1968] using the 3.39 μm line of a He–Ne laser with CH_4 as the absorber. Malyshev *et al.* [1980] have reported that this line is reproducible to 5 parts in 10^{13} and its frequency has been measured with an accuracy of ± 3 parts in 10^{11} [Knight *et al.*, 1980].

In the visible region, both $^{127}I_2$ and $^{129}I_2$ have absorption lines that coincide with the He–Ne laser line at 633 nm [Hanes and Dahlstrom, 1969; Knox and Pao, 1970]. Because the saturated absorption peaks in iodine are very small ($< 1\%$) and are superposed on a sloping background, it is necessary to use a third-harmonic servo system to eliminate systematic frequency shifts due to this slope [Wallard, 1972; Hanes *et al.*, 1973]. A detailed description of such an iodine-stabilized He–Ne laser operating at a wavelength of 633 nm has been given by Layer [1980]. This laser has an output whose frequency is stable to 3 parts in 10^{13} over 1000 s and reproducible to better than 5 parts in 10^{10}.

This technique of frequency stabilization has been the subject of intensive research [Brillet and Cérez, 1981] and lasers using it now provide a number of wavelengths which have replaced the ^{86}Kr lamp for interferometric measurements of length (see Section 8.8).

6.4.5 Stabilization by saturated fluorescence

The most common infrared laser is the CO_2 laser. A widely used method of frequency stabilization for this laser involves an interesting variation of the technique of saturated absorption. Low-pressure CO_2 in an auxiliary cell absorbs weakly at 10.6 μm, because the lower laser level has a small thermal population, and, as a result, fluoresces in the 4.3 μm band which connects the upper laser level to the ground state. Due to saturation of the absorption, the fluorescence efficiency exhibits a dip at line centre. If the laser is frequency modulated, the resultant signal can be used very effectively to stabilize the laser centre frequency [Freed and Javan, 1970]. This technique has the advantage that it permits stabilization on any one of a number of transitions in the 10.6 μm band on which the CO_2 laser can oscillate. A frequency stability of 1 part in 10^{12} can be obtained, with a reproducibility of about 2 parts in 10^{10}.

6.5. Semiconductor lasers

Semiconductor lasers are now being used to an increasing extent in interferometry. The first semiconductor lasers [Hall *et al.*, 1962; Nathan *et al.*, 1962; Quist *et al*, 1962; Holonyak and Bevacqua, 1962] consisted essentially of a small rectangular chip of a suitable semiconductor (GaAs) in which a planar $p-n$ junction had been produced. When a current is passed through this junction, minority carriers (mainly electrons) are injected across the $p-n$ junction into a region typically about 10 μm thick where they recombine with the majority carriers. This normally results in spontaneous emission at a wavelength corresponding to the energy gap between the valence and conduction bands (λ = 840 to 910 nm, depending on the temperature, for GaAs). However, when the current density is raised to a sufficiently high level, a population inversion is produced, resulting in stimulated emission at this wavelength. If the ends of the chip are cleaved to produce smooth, parallel faces, their reflectance (about 30%) is adequate, because of the very high gain of the junction layer, to obtain laser oscillation.

A major disadvantage of such a simple $p-n$ junction semiconductor laser is that the threshold current for laser oscillation is very high, unless the chip is cooled to well below room temperature. However, the threshold current can be reduced by having, on each side of the GaAs inversion layer, an additional layer of a semiconductor with a higher band gap. Such multilayer structures are commonly called heterostructures. For good performance, the crystal lattice dimensions of the two materials must be very similar. This condition is satisfied by $Ga_{1-x}Al_xAs$ layers, and heterostructures fabricated with the GaAs/(GaAl)As system have made possible semiconductor lasers which operate continuously at room temperature [Kressel and Nelson, 1969; Hayashi et al., 1969; Panish et al., 1970]

In addition, the operating current has to be confined to a narrow stripe along the axis of the laser to limit the sideways spread of injected carriers and the consequent increase in the width of the region of laser emission in the plane parallel to the junction. This can be done in different ways, resulting in a number of configurations known as stripe-geometry structures. Various types of heterostructures and stripe geometries have been developed to optimize different characteristics [Thompson, 1980]. Figure 6.7 shows a simple design using two parallel insulating layers of SiO_2, one on either side of the stripe contact [Ripper et al., 1971].

Semiconductor lasers have the unique advantages of small size, simplicity of operation and high efficiency. However, because of the small size of the active region, it is difficult to control the mode structure, and the output beam divergence is quite large, typically $5° \times 50°$. In addition, since stimulated emission takes place between the edges of two broad bands, the spectral bandwidth over which the laser emits, even when the transverse modes are suppressed, is usually much larger than in a gas laser. Thus, at threshold, the output may consist of ten or more longitudinal modes with a frequency separation of about 80 GHz, though with a temperature-stabilized laser, the width of individual modes may be less than 40 MHz. However,

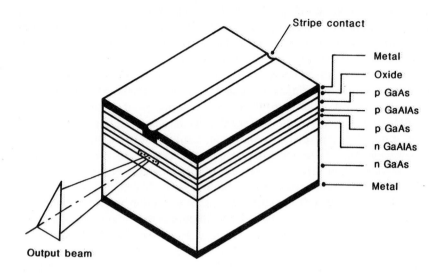

Stripe contact

Metal
Oxide
p GaAs
p GaAlAs
p GaAs
n GaAlAs
n GaAs
Metal

Output beam

Fig. 6.7. Stripe-geometry double-heterostructure junction laser. Typical dimensions: 400 μm long, 250 μm wide and 120 μm thick; stripe width 15 μm [Ripper *et al.*, 1971].

with proper design, the number of longitudinal modes decreases rapidly as the current is increased, permitting single-frequency operation [Streifer *et al.*, 1977; Aiki *et al*, 1978; Kajimura *et al.*, 1979]. A narrower line width as well as a more stable output frequency can be obtained by antireflection coating the end faces of the chip and operating it in an external cavity with two low-finesse Fabry-Perot étalons as wavelength-selecting elements [Voumard, 1977].

6.6 Ruby and Nd:YAG lasers

A laser can also be built using as the active medium a solid rod of a suitably doped crystal or glass. In this case, the population inversion is produced by irradiating the rod with light from a flashlamp, and the laser output also takes the form of a pulse or a series of pulses lasting a few microseconds. The most commonly used material is ruby (λ = 694 nm). An alternative is neodymium-doped yttrium aluminium garnet (Nd:YAG) (λ = 1.06 μm), the output from which can be converted to visible light by means of a frequency-doubling crystal [see Koechner, 1976].

Very short pulses of light ($\approx$ 15 ns duration) can be produced by using a Q switch in the laser cavity. This is a fast optical shutter, typically a Pockels cell, which is normally closed so that the laser cannot oscillate when the flash lamp is fired. As a result, a large population inversion can be built up and a short pulse with very high peak power is produced when the switch opens.

Because of the high gain of these materials, these lasers normally oscillate in a number of modes, so that the coherence of the output is relatively poor. However, where an extended coherence length is needed, operation in a single longitudinal mode can be ensured by using a mode-selecting aperture and a Fabry-Perot étalon in the cavity.

6.7. Dye lasers

The use of a solution of a suitable dye as the active medium in a laser has several advantages: low cost, high gain and a tunable output.

Initially, most dye lasers were built for pulsed operation, but it was soon shown by Peterson *et al.* [1970] that the pump-power density requirements for continuous-wave (cw) operation could be obtained by focusing a high-power Ar^+ laser on the flowing dye.

Because of the very large number of dyes available, laser operation is possible at a number of wavelengths [see, for example, the detailed review by Schafer, 1973]. In addition, the output from a given dye can be tuned continuously over a range of wavelengths (typically about 70 nm), by incorporating a wavelength selective element in the resonant cavity (see Section 10.7). With a suitable design [Jarrett and Young, 1979], tunable single-frequency radiation at power levels up to a few hundred milliwatts can be obtained over the entire visible spectrum.

6.8 Laser beams

The beam from a laser has a Gaussian intensity profile and its diameter is often much smaller than the aperture of the interferometer. It is then necessary to expand the beam to obtain a reasonably uniform intensity distribution over the field. This can be done by a microscope objective which brings the beam to a focus, followed, where a collimated beam is required, by a lens of appropriate focal length. The expanded beam may exhibit diffraction patterns (spatial noise) due to dust on the surfaces of the microscope objective, but these can be eliminated by a pinhole placed at its focus which acts as a spatial filter and blocks the diffracted light. If a pinhole is chosen whose diameter is slightly greater than the central maximum of the image (the Airy disc), the loss of light is negligible.

Laser light can also give rise to other practical problems because of its high spatial and temporal coherence. If a diffuser is introduced in the beam to obtain an extended diffuse source, speckle (see Chapter 14) can be a nuisance. However, it can be eliminated when observing or photographing the fringes by moving the diffuser continuously in its own plane. A more serious problem is reflected light from the various surfaces in the optical paths in the interferometer, which can give rise to spurious fringe systems. These are best eliminated by properly oriented polarizers; alternatively, stops located at suitable points in the system can be used. It is also necessary to avoid feedback due to light reflected back from the interferometer into the laser, since changes in the amplitude or the phase of the reflected light can cause changes in the output or even (see section 6.4.1) the frequency of the laser.

Finally, it is very important when working with lasers to see that adequate precautions are taken to avoid damage to the retina, including the use of protective eyewear where necessary (see ANSI Standard Z 136.1-1980, American National Standard for the Safe Use of Lasers).

7
Measurements of length

One of the main applications of interferometry is in accurate measurements of length. This essentially involves the determination of the mean interference order over a specified area of the interference pattern. Such length measurements are usually carried out on material standards of two types: line standards, consisting of a set of fine lines engraved on a polished surface, and end standards, consisting of a bar with polished, flat, parallel ends.

7.1. Line standards

Measurements on line standards, such as scales, are carried out with an interference comparator [see the review by Baird, 1963]. As shown schematically in Fig. 7.1, the scale to be calibrated is mounted on a movable carriage which also carries one of the mirrors of a Michelson interferometer. Successive graduations on the scale are located by a fixed photoelectric microscope, while the displacement of the mirror is evaluated from the corresponding change in the interference order. A transfer mirror provides a reference surface which can be used to measure a 1 m scale in two stages, when working with radiation of limited coherence length.

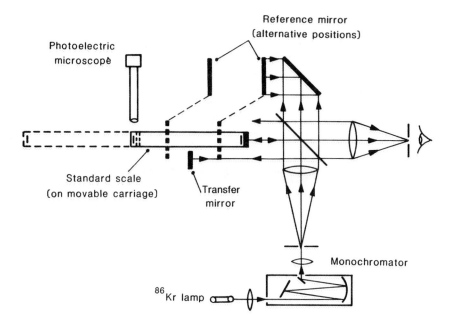

Fig. 7.1. Schematic diagram of an interference comparator for 1-metre line standards [Ciddor and Bruce, 1967].

7.2. End standards

Measurements of short end standards can be made with a Fizeau interferometer, but longer end standards are usually calibrated with a Kösters interferometer. As shown in Fig. 7.2, this is basically a Michelson interferometer using collimated light which incorporates a dispersing prism to select a single spectral line from the source. The end standard is wrung on to a metal optical flat and the difference in the interference orders for the flat and the top of the end standard is determined. The Kösters interferometer has the advantage over the Fizeau interferometer that the image of the reference mirror can be positioned halfway between the faces of the end standard, so that the maximum path difference is less than half that in the Fizeau interferometer.

7.3. Determination of the integral order

A problem in all interferometric measurements of length is that if the interference order with monochromatic light of wavelength λ is, say, $(m+\epsilon)$,

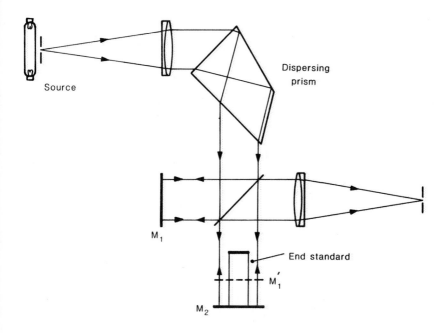

Fig. 7.2. Kösters interferometer for end standards.

where m is an integer and ϵ is a fraction, observations on the fringes only yield the value of ϵ. Various methods have been used to find m, the integral order. The most direct method is by counting the fringes which pass a given point in the field while one of the mirrors of the interferometer is moved over the distance to be measured; this method can be used very effectively with photoelectric fringe counting techniques (see Chapter 9). Another method involves the use of the zero-order white light fringe to judge the equality of the two optical paths in the interferometer; this can be used when comparing two end standards of very nearly the same length. Yet another method is optical multiplication. If two Fabry-Perot interferometers are placed in series, the spacings of the plates in the two interferometers can be adjusted, using fringes of superposition (see Section 4.5), so that they are in an integral ratio. This permits measuring a large optical path in terms of a shorter known optical path. However, the most commonly used method is the method of exact fractions.

7.4. The method of exact fractions

In this method, the interferometer is illuminated successively with different wavelengths $\lambda_1, \lambda_2, \ldots \lambda_n$, and the fractional orders of interference $\epsilon_1, \epsilon_2, \ldots \epsilon_n$

are measured for these wavelengths. If p is the optical path difference, we then have

$$p = (m_1 + \epsilon_1)\lambda_1 \qquad (7.1)$$

$$p = (m_2 + \epsilon_2)\lambda_2 \qquad (7.2)$$

$$\dots\dots\dots\dots\dots$$

$$p = (m_n + \epsilon_n)\lambda_n \qquad (7.3)$$

where $m_1, m_2, \dots m_n$ are the corresponding integral orders. Since the value of p is already known within fairly close limits from mechanical measurements, an approximate estimate of the integral order can be made for one of the wavelengths, say λ_1. A range of integral values for m_1, centred on this value, is then taken, and the corresponding values of $\epsilon_2 \dots \epsilon_n$ are then calculated from Eqs (7.1) to (7.3). The correct value of m_1 is then that for which the calculated values of all the excess fractions agree with the measured values to within the experimental uncertainty.

The method of exact fractions can be applied very effectively with a CO_2 laser, since a number of lines whose wavelengths are known with a high degree of accuracy are available with such a source [Bourdet and Orszag, 1979; Gillard et al., 1981; Gillard and Buholz, 1983]. A set of analytical equations amenable to automatic computation has been developed by Tilford [1977] that permit the length to be calculated by a series of approximations in terms of the wavelengths and the measured fractional interference orders. These equations can also be used to choose suitable wavelengths for the range to be covered.

7.5. Measurement of refractive index

A problem closely related to the measurement of length is the measurement of the refractive index of air. Unless a length measurement is carried out in a vacuum, the value obtained for the optical path length has to be divided by the refractive index of the air within the interferometer to obtain the true length.

Detailed measurements of the refractive index of air were made by Barrell and Sears [1939] using two Fabry-Perot interferometers of almost the same length in series. To start with, the shorter one was filled with air while the other was evacuated. The first interferometer was then slowly evacuated and the change in the interference order was measured from observations

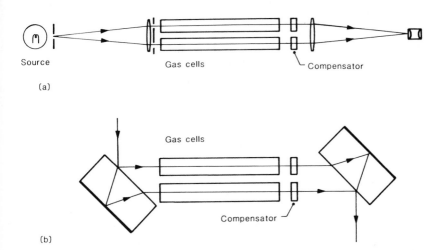

Source   Gas cells   Compensator

(a)

(b)

Fig. 7.3.   (a) Rayleigh, and (b) Jamin interferometers.

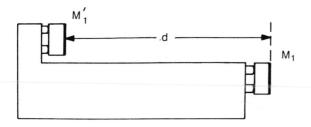

Fig. 7.4.   Transfer standard used by Michelson.

on the fringes of superposition. These values have been used in formulas developed by Edlén [1966] and by Jones [1981] which can be applied to estimate the refractive index correction, provided the air is free from contamination. However, where the highest precision is needed, an evacuated cell inserted in one of the beams is used to determine the actual refractive index of the air within the instrument.

Interferometry has also been widely used to measure the refractive indices of other gases, as well as mixtures of gases, since they are all so close to unity that measurements by conventional methods of refractometry are not possible. Two interferometers commonly used for this purpose are the Rayleigh interferometer and the Jamin interferometer. As shown in Fig. 7.3, the two beams pass through two cells of the same length containing the gases whose refractive indices are to be compared. For absolute measurements of the refractive index, one cell is evacuated and the number of fringes crossing a given point in the field is counted as the gas is admitted to the cell. Since this can be rather tedious, a compensator is frequently used. This consists of a fixed glass plate in one beam, and an identical plate in the other beam which can be tilted to introduce a known change in the optical path. However, it should be noted that, as discussed in Section 3.3, such measurements only give the phase index when the actual interference order is measured with a quasi-monochromatic source. Measurements which involve judgements of equality of the two optical paths using white-light fringes give values of the group index.

7.6. Measurements of the International Prototype Metre

The first experimental measurement of the length of the Pt–Ir bar which was the international prototype of the metre, in wavelengths of light, was carried out by Michelson in 1892, using the red cadmium line ($\lambda \approx 644$ nm) and a set of nine transfer standards. Each of these consisted, as shown in Fig. 7.4, of two mirrors $M_1 M_1'$ set parallel to each other on a metal support. The distance d separating the mirrors increased by a factor of 2 from approximately 0.39 mm in the shortest (Standard 1) to 100 mm in the longest (Standard 9).

The first step was to measure the separation of the mirrors M_1, M_1' in Standard 1 in terms of the wavelength of the red cadmium line. For this, Standard 1 was set up in an interferometer as shown in Fig. 7.5 (a). The half of the field covered by M_1 and M_1' was illuminated with white light, while the other half was illuminated with monochromatic light from the cadmium lamp. To start with, the end mirror M_B of the interferometer was set so that its image M_B' coincided with M_1. This was done by locating the zero-order fringe at the center of M_1. The end mirror M_B was then moved slowly until M_B' coincided with M_1'. During this operation the number of fringes passing a given point in the interference pattern formed in the other half of the field by M_B and M_A was counted. Since the distance through which M_B had been moved was equal to the separation of the mirrors in Standard 1, this count gave the length of Standard 1 in terms of the wavelength of the red cadmium line.

The next step was to compare Standard 1 with Standard 2. For this, the two standards $M_1 M_1'$ and $M_2 M_2'$ were set up, side by side, in one arm of the interferometer, as shown in Fig. 7.5 (b), and adjusted so that M_B' coincided with M_2' as well as M_1'. The end mirror M_B was then moved so that M_B' coincided with M_1. After this, Standard 1 was moved so that M_1' coincided

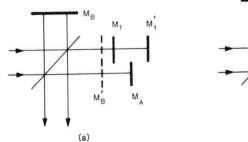
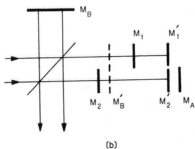

(a)   (b)

Fig. 7.5. (a) Measurement of Transfer Standard 1 in wavelengths, and (b) comparison of Transfer Standard 1 with Transfer Standard 2.

with M'_B, and M_B was moved so that M'_B coincided once again with M_1. As a result of these three steps, M_B was moved through a distance exactly equal to twice the length of Standard 1. To measure the small distance between this position of M'_B and M_2, Standard 1 was removed, and the number of monochromatic-light fringes crossing this half of the field as M'_B was moved into coincidence with M_2 was counted. This process was repeated with successive standards until the length of Standard 9 was determined.

The final step was to compare Standard 9 with the prototype metre. For this, the standard was placed on the carriage of the interferometer by the side of the prototype metre, and a microscope was set up over each of the terminal lines on the metre. A fine line engraved on the edge of the standard was then brought up to one end of the end lines on the metre, and the distance between the two lines was read with the microscope. After this, Standard 9 was moved in steps, using white-light fringes as described above, through a distance equal to 10 times its length. The short distance separating the line on Standard 9 and the line at the other end of the metre was then read off the second microscope.

Another series of measurements was carried out by Benoit, Fabry and Perot in 1913 using multiple-beam fringes formed between pairs of plane, semi-transparent mirrors. Five étalons were used with u-shaped invar separators, their lengths being 0.0625m, 0.125m, 0.25m, 0.5m, and 1 m. The last étalon had two fine lines engraved on the upper edges of the mirrors, parallel to their surfaces.

The measurements, in this case also, involved three steps. In the first step, the distance between the two marks on the 1 m étalon was determined with reference to the prototype metre, using an optical comparator. The next step was to determine N_0, the sum of the distances between the lines and the faces of the mirrors, expressed in wavelengths. This was done in a separate experiment in which the plates were mounted to form, in turn, two short étalons in one of which the distance between the lines was twice that in the other. If N_1 and N_2 are the orders of interference with these two short étalons, $N_0 = 2N_2 - N_1$

The final step involved measuring, in wavelengths, the distance between the faces of the plates in the 1 m étalon. Since it was not possible to obtain fringes with such a large optical path with the cadmium lamp, it was necessary to start by measuring the separation of the plates in the 6.25 cm étalon. This was done by the method of exact fractions. This étalon was then compared with the next longer one using fringes of superposition, the residual difference being measured, as shown in Fig. 7.6, by means of a third, slightly wedged air film, which could be moved across the field to introduce an additional optical path and obtain white-light fringes.

7.7. The ^{86}Kr standard

The measurements described in Section 7.6, as well as subsequent measurements at a number of national laboratories and the *Bureau International des Poids et Mésures* (BIPM), showed that while the definition of the metre based on the international prototype was significant to about 0.2 μm, or 2 parts in 10^7, lengths could be measured to a much higher degree of precision in terms of the wavelength of a suitable spectral line. This resulted in a considerable body of opinion that favoured replacement of the prototype metre by a spectral line. Finally, after detailed studies of the characteristics of the spectral lines of ^{114}Cd, ^{198}Hg and ^{86}Kr, the metre was redefined in 1960 in terms of the wavelength of the orange line of ^{86}Kr (λ = 606 nm). Because ^{86}Kr has no nuclear magnetic moment, this line exhibits no hyperfine structure and has a symmetrical profile. In addition, the broadening of the line due to the Doppler effect can be minimized by operating the source in a bath of nitrogen at its triple point (64 K). Measurements with this source are reproducible to better than 1 part in 10^8 [Baird and Howlett, 1963].

7.8. Frequency measurements and the speed of light

With the development of stabilized lasers (see Section 6.3) sources of monochromatic light became available with much narrower linewidths and much better reproducibility than the ^{86}Kr lamp. The wavelength of the 3.39 μm line from the He–Ne laser stabilized by saturated absorption in CH_4 was the first to be measured with high accuracy relative to the ^{86}Kr standard by several national laboratories and the BIPM. These measurements had a spread of only 6 parts in 10^9, which was largely attributable to the uncertainty associated with the ^{86}Kr standard itself.

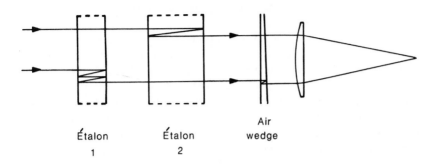

Air

Étalon   Étalon   wedge

1   2

Fig. 7.6. Comparison of two étalons using fringes of superposition.

Simultaneously, the major national laboratories undertook measurements to compare the frequency of the same laser with the frequency of the ^{133}Cs clock which is the primary standard of time. This comparison necessarily had to be carried out in stages. In one experiment [Evenson *et al.*, 1972, 1973] the first stage involved a klystron oscillating at 74 GHz whose frequency could be compared directly with the ^{133}Cs standard ($\nu \approx$ 9.19 GHz). The twelfth harmonic of this klystron gave a beat frequency of about 2 GHz with an HCN laser ($\nu \approx$ 890 GHz), which could be measured precisely using a diode detector consisting of a tungsten wire a few micrometres in diameter with a sharpened end contacting a polished nickel surface. Measurements were made similarly of the beat frequency between the twelfth harmonic of the HCN laser and a H_2O laser ($\nu \approx$ 10.72 THz). The third harmonic of this laser gave a measurable beat with a CO_2 laser ($\nu \approx$ 32.13 THz). This laser was, in turn compared with a CO_2 laser oscillating on another transition ($\nu \approx$ 29.44 THz), whose third harmonic gave a beat with the He–Ne laser ($\nu \approx$ 88.38 THz). It was possible, in this fashion, to measure the frequency of the He–Ne laser with an uncertainty of 6 parts in 10^{10}.

A simpler setup for such a frequency chain using beats between five CO_2 lasers with different isotope fillings operating on slightly different wavelengths has been described by Whitford [1979], while the extension of frequency measurements to the visible region has been discussed by Evenson *et al.* [1981] and by Baird [1981, 1983a].

Since the speed of light is given by the product of the wavelength of a source and its frequency, these results give a value for the speed of light. The accuracy of this value

$$c = 299\ 792\ 458\ \text{ms}^{-1} \tag{7.4}$$

which was formally adopted in 1975, is limited essentially by the uncertainties relating to the standard of length, the wavelength of the ^{86}Kr lamp.

7.9. The definition of the metre

It is apparent from Section 7.8 that a point had been reached at which the ^{86}Kr standard was no longer adequate for absolute measurements of the wavelengths of stabilized lasers with the degree of precision justified by their characteristics; an improved standard of length was necessary. An obvious possibility was to replace the ^{86}Kr lamp by a stabilized laser. However, this could lead to the need for further revisions each time a better laser source was developed. After careful consideration of various alternatives, the *Comité Consultatif pour la Définition du Mètre* (CCDM) therefore proposed a different solution which has many advantages in the long term [Terrien, 1976; see also the review by Petley, 1983]. This was to ascribe a fixed value, based on the best results obtained so far, to the speed of light, which is a physical constant. Specified laser radiations would then be involved in the methods of realization of the metre, rather than in its definition.

Accordingly, from October 1983, the metre has been redefined as follows:

> The metre is the length of the path travelled by light in vacuum during a,time interval of 1/299 792 458 of a second.

Three distinct methods of realizing the metre have been recommended by the CCDM. The first is by direct measurements of the travel time of an electromagnetic wave. The second is by measurements in terms of the wavelength in vacuum of an electromagnetic wave of known frequency. The third is by means of the vacuum wavelengths of a number of specified sources; these include lasers stabilized by saturated absorption (reproducibility between 2 parts in 10^9 and 2 parts in 10^{10}) as well as spectral lamps using ^{86}Kr, ^{198}Hg and ^{114}Cd (reproducibility between 1 part in 10^8 and 1 part in 10^9).

7.10. The speed of light and the Michelson–Morley experiment

The definition of the metre discussed in Section 7.9 assumes that the speed of light is a constant. This is an axiom in the theory of relativity, which is now well-established, but it was not apparent to scientists of the nineteenth century who believed that light was a transverse wave, propagated through an all-pervading medium called the 'luminiferous aether'. A question which arose, naturally, was what effect the movement of the earth through this medium had on the speed of light. A very important experiment using an interferometer carried out by Michelson in 1881 and repeated, in collaboration with Morley, in 1887, was designed to answer this question.

Consider an interferometer in which, as shown in Fig. 7.7(a), the arm OM_1 is parallel to the direction in which the earth is moving with a velocity

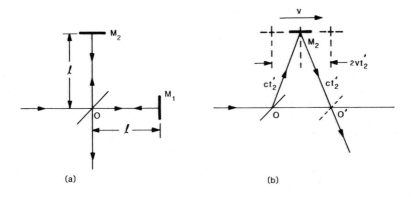

Fig. 7.7. Theory of the Michelson–Morley experiment.

v through space. According to the classical laws of physics, the time taken for a light wave to travel from O to M_1 would be

$$\tau_1' = l/(c-v) \tag{7.5}$$

where l is the length of the arm, while the time taken for the light wave to travel from M_1 to O would be

$$\tau_1'' = l/(c+v) \tag{7.6}$$

Accordingly, the total delay for the path OM_1O would be

$$\begin{aligned}
\tau_1 &= \tau_1' + \tau_1'' \\
&= 2l/c(1 - v^2/c^2) \\
&\approx (2l/c)\,(1 + v^2/c^2)
\end{aligned} \tag{7.7}$$

since $v \ll c$

To calculate τ_2, the delay for the path OM_2, we have to allow for the lateral displacement of O during this time. We then have, from Fig. 7.7(b),

$$c^2 \tau_2'^2 = l^2 + v^2 \tau_2'^2 \tag{7.8}$$

where τ_2' is the time taken to cover the path OM_2. This is also the time taken to cover the path M_2O'. The total delay for this path is, therefore,

$$\begin{aligned}
\tau_2 &= 2\tau_2' \\
&= (2l/c)\,(1 - v^2/c^2)^{-1/2} \\
&\approx (2l/c)\,(1 + v^2/c^2)
\end{aligned} \tag{7.9}$$

It follows that, even if the two paths are equal there should be a difference in the delays

$$\Delta\tau = \tau_1 - \tau_2$$
$$= (l/c)(v/c)^2 \qquad (7.10)$$

so that a change in the delay of $2\Delta\tau$ should be observed if the interferometer is rotated through $90°$; this effectively shifts the additional delay from one path to the other. The absence of any such change in the delay led to the rejection of the concept of the aether [Shankland, 1973].

The Michelson–Morley experiment was repeated several times between 1905 and 1930, but always with a null result. This null result has also been confirmed to a much higher degree of accuracy using laser heterodyne techniques [Jaseja et al., 1964; Brillet and Hall, 1979]. In the latter experiment, the frequency of a He–Ne laser ($\lambda \approx 3.39\ \mu m$) was locked to the resonant frequency of a very stable, thermally isolated Fabry–Perot interferometer which was mounted along with it on a rotating horizontal granite slab. As a result, any variations in the delay within this Fabry–Perot interferometer would appear as variations of the laser frequency. Such variations could be detected with an extremely high degree of sensitivity by measuring the beat frequency generated by mixing this beam (see Section 6.2) with the beam from a stationary reference laser which was stabilized by saturated absorption in CH_4. However, no changes in the beat frequency greater than 5×10^{-7} of that predicted by classical theory were found.

7.11. Interferometric measurements of rotation

Interferometric experiments designed to detect rotation are in a different category, since both classical theory and the theory of relativity predict a fringe shift [see the reviews by Post, 1967; Hariharan, 1975c]. Sagnac demonstrated in 1913 the feasibility of such an experiment with an interferometer of the form shown in Fig. 2.14(b), in which interference takes place between two beams travelling around the same circuit in opposite directions. When the whole interferometer, including the light source, was set rotating with an angular velocity ω about an axis making an angle θ with the normal to the plane of the interferometer, a fringe shift was observed corresponding to the introduction of a delay between the two beams

$$\tau = 4\omega A \cos\theta/c^2 \qquad (7.11)$$

where A was the area enclosed by the light path. The fringe shift given by Eq. (7.11) could be doubled by comparing the fringe positions for rotation at the same speed in opposite directions. Sagnac also established that the effect does not depend on the shape of the loop or the position of the centre of rotation.

Subsequently, in 1925, Michelson and Gale succeeded in demonstrating the rotation of the Earth by means of a similar experiment. To obtain the necessary sensitivity they used a rectangular path 60 m × 15 m enclosed in evacuated tubes. Since the rate of rotation could not be changed, the fringe shift was measured by comparing the fringes in the main circuit with the fringes formed in a comparison circuit enclosing a smaller area.

Much higher sensitivity can be obtained by measurements of the optical beat frequency in a ring laser [Macek and Davis, 1963], such as that shown schematically in Fig. 7.8. When such a laser is rotated, the clockwise-propagating (CW) and counter-clockwise-propagating (CCW) modes have their frequencies shifted by equal amounts in opposite senses. A very stable beat can be obtained with such a laser, because the two modes use the same optical cavity and are affected equally by any mechanical instabilities.

Due to the presence of the active medium within the ring, the two modes in such a ring laser are prone to locking at low rotation rates. Some solutions to this problem have been discussed by Roland and Agrawal [1981]. Alternatively, it can be eliminated by using a passive ring interferometer with an external laser, and measuring the differential shift in resonant frequency for the two directions of propagation [Ezekiel and Balsamo, 1977]. One scheme uses a single laser and two acousto-optic modulators to generate two independently controlled optical frequencies which are locked to the CW and CCW resonant frequencies of the ring. Another uses only one laser frequency and a Faraday cell within the cavity to cancel out any difference in the optical path lengths for the counter-propagating beams. Yet another approach has been to use a multi-turn fibre-optic loop instead of a conventional cavity with mirrors to increase the effective area of the ring [Vali and Shorthill, 1976; Leeb et al., 1979]. The relative merits of these and other techniques have been reviewed in a number of papers [see, for example, Ezekiel and Knausenberger, 1978; Ezekiel and Arditty, 1982]. Ring laser rotation sensors have the advantages of fast warm-up, rapid response, large dynamic range, insensitivity to linear motion and freedom from cross-coupling errors in multi-axis sensing. For these reasons they are now widely used in missile guidance and intertial navigation systems.

7.12. Gravitational waves

Estimates by astrophysicists suggest that the detection of gravitational waves requires a sensitivity to strains of the order of a few parts in 10^{-21}.

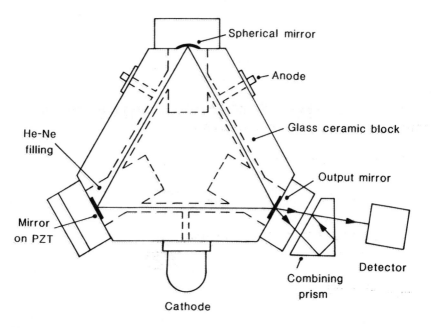

Fig. 7.8. Triangular ring laser for rotation sensing [Roland and Agrawal, 1981].

Techniques which have been proposed for this purpose include high precision laser interferometry [Maischberger *et al.*, 1981].

One type of detector which is under construction uses two identical
Fabry–Perot interferometers, about 1000 m long, at right angles to each
other [Drever *et al.*, 1981], the mirrors of these interferometers being mounted on four nearly free test masses. This detector makes use of the fact that
a gravitational wave travelling in a suitable direction will affect the two
baselines in opposite senses. The separations of the mirrors are continuously
compared by servo-locking the output frequency of a laser to the transmission peak of one interferometer, while the optical path in the other is
continuously adjusted by a second servo system so that its transmission peak
also coincides with the laser frequency. The corrections applied to the second
interferometer then provide a signal which can be used to detect any changes
in the length of one arm with respect to the other, over a wide frequency
range, with extremely high sensitivity.

8

The study of optical wavefronts

Interferometry can be used to study phase variations across an optical wavefront. As mentioned in Chapter 2, one field of application has been in studies of gas flows, combustion and diffusion, where local changes in the refractive index can be related to changes in the pressure, the temperature or the relative concentration of different components. Another application has been in microscopy, for the study of surface relief or structures in transparent specimens. Finally, a very important application is in testing optical components and optical systems.

8.1. The Fizeau interferometer

Fringes of equal thickness (see Section 2.4) can be used to compare a nominally flat surface with a reference optical flat. The simplest method of doing this is to bring the two surfaces together and view the fringes formed in the thin air film separating them at near normal incidence. If the deviations of the surface from flatness are small, the top plate is tilted slightly to obtain a wedged air film. If, then, the fringe spacing is a and the peak deviation of the diametral fringe from straightness is δ, the peak error of the surface is $(\delta/a) (\lambda/2)$.

However, as shown in Section 2.4.3, fringes of equal thickness are obtained with an extended source only when the air gap between the surfaces is less than a few micrometres. To obtain fringes of equal thickness with a larger

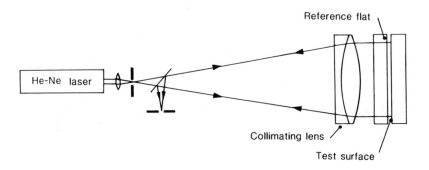

Fig. 8.1. Optical system of a Fizeau interferometer.

air gap, it is necessary to use collimated light. Figure 8.1 is a schematic diagram of a Fizeau interferometer using a lens for collimation; alternatively, a concave mirror can be used. Increased accuracy can be obtained if reflecting coatings are applied to the reference flat and the surface under test, so that multiple-beam fringes (see Section 4.3) are formed [Koppelmann and Krebs, 1961, a,b].

With such an interferometer, it is possible, provided care is taken to exclude vibrations, to use a liquid surface as a reference flat [Bünnagel, 1956, 1965]. Alternatively, three surfaces can be tested in a series of combinations to evaluate their individual errors [Schulz and Schwider, 1967, 1976].

One common application of the Fizeau interferometer is to check the parallelism of the faces of a transparent plate. With a laser source, problems of coherence are eliminated and quite thick plates can be tested. Modified forms of the Fizeau interferometer have also been described using convergent or divergent light to test curved surfaces [Heintze *et al.*, 1967; Biddles, 1969; Shack and Hopkins, 1979].

8.2. The Twyman–Green interferometer

A much wider range of optical components can be tested with the Twyman–Green interferometer [see Twyman, 1957]. This is a Michelson interferometer modified to use collimated light, so that fringes of equal thickness are obtained. Detailed descriptions of a number of test setups for various optical elements have been given by Briers [1972] and by Malacara [1978a]. The optical arrangement used to test a prism is shown in Fig. 8.2(a), while that used to test a lens is shown in Fig. 8.2(b). With the addition of a nodal slide, lenses can also be tested off-axis.

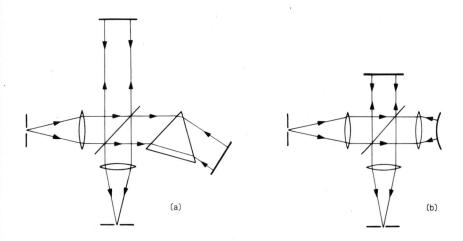

Fig. 8.2. Twyman–Green interferometer used to test (a) a prism, and (b) a lens.

For a lens used off-axis, the deviation of the transmitted wavefront from a sphere centred on the Gaussian image point can be written as

$$W(x,y) = A(x^2 + y^2)^2 + By(x^2 + y^2) + C(x^2 + 3y^2)$$
$$+ D(x^2 + y^2) + Ey + Fx \qquad (8.1)$$

where A is the coefficient of spherical aberration, B the coma coefficient, C the astigmatism coefficient, D the defocusing coefficient and E and F are measures of tilt about the x and y axes, respectively [Kingslake, 1925-1926].

Pictures of interferograms corresponding to the various primary aberrations have been presented by Kingslake [1925-1926] and by Malacara [1978a]. An interferogram containing a mixture of aberrations can be evaluated by measuring the optical path difference at several points on the x and y axes and then calculating the coefficients in Eq. (8.1) from a set of linear equations. This process can be simplified by double-passing the interferometer so that the symmetrical and antisymmetrical wavefront aberrations are displayed separately [Hariharan and Sen 1961d]. Computerized analysis is also possible with a least-squares program, which can be used with wavefronts of arbitrary shape [Rimmer et al., 1972; Freniere et al., 1981].

To obtain fringes of good contrast with a thermal source, its size and spectral bandwidth must satisfy the conditions for spatial and temporal coherence (see Sections 3.10 and 3.11). The conditions for this have been derived by Hansen [1955] and by Hansen and Kinder [1958]. However, with the use of a laser source, these problems disappear, and fringes can be

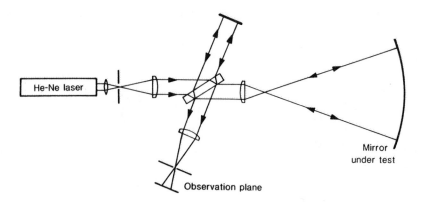

Fig. 8.3. Laser unequal-path interferometer used for testing a large concave mirror [Houston *et al.*, 1967].

obtained even with large path differences between the beams. Figure 8.3 shows a versatile unequal path interferometer which can be used for testing large concave mirrors [Houston *et al.*, 1967].

8.3. Shearing interferometers

In a shearing interferometer, both the wavefronts are derived from the system under test, and the interference pattern is produced by shearing one wavefront with respect to the other. Shearing interferometers have the advantage that they do not require a reference surface of the same dimensions as the system under test. In addition, since both the interfering beams traverse very nearly the same optical path, the fringe pattern is much less affected by mechanical disturbances.

To produce fringes of good visibility with a shearing interferometer, the beams must have adequate spatial coherence. With a thermal source, this requires a very small illuminated pinhole and, hence, an intense source, usually giving light of poor temporal coherence. As a result, much effort went initially into the design of shearing interferometers which were compensated for white light [see the review by Bryngdahl, 1965]. With the availability of gas lasers which give light with a high degree of spatial and temporal coherence, many simpler arrangements have come into use.

8.3.1. Lateral shear

The simplest type of shear is a lateral shear (see Section 3.9). With a nearly plane wavefront, this involves producing two images of the wavefront with

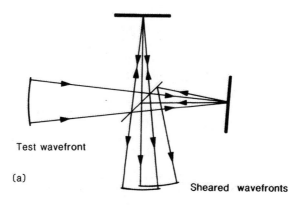

Test wavefront

(a)

Sheared wavefronts

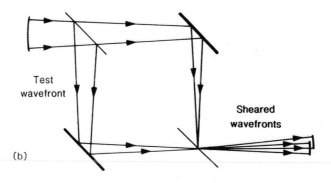

Test
wavefront

Sheared
wavefronts

(b)

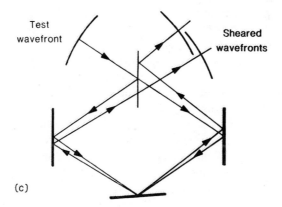

Test
wavefront

Sheared
wavefronts

(c)

Fig. 8.4. Lateral shearing interferometers based on (a) the Michelson, (b) the Mach–Zehnder, and (c) the Sagnac interferometers.

a small mutual lateral displacement, while with a nearly spherical wavefront, it requires a similar displacement of the two images of the wavefront over the surface of the reference sphere. Figure 8.4 shows three typical optical systems which can be used with converging wavefronts. These systems, which are based on the Michelson, Mach–Zehnder and Sagnac interferometers, were described by Lenouvel and Lenouvel [1938], Bates [1947], and Hariharan and Sen [1960e] respectively. A number of modifications of these for use with spherical and flat wavefronts have been reviewed by Murty [1978].

A particularly simple arrangement devised by Murty [1964] can be used with a laser source and consists of a plane-parallel plate. As shown in Fig. 8.5, the light from the laser is focused by a microscope objective on a pinhole located at the focus of the lens under test. The beam emerging from this lens gives rise to two wavefronts reflected from the front and back surfaces of the plate. The lateral shear between these wavefronts can be varied by tilting the plate and is given by the relation

$$s = (n^2 - \sin^2\theta)^{-\frac{1}{2}} \, d \sin 2\theta \tag{8.2}$$

where d is the thickness of the plate and θ is the angle of incidence. A modification of this arrangement [Hariharan, 1975b] uses two separate plates with a variable air gap; this has the advantage that a tilt can be introduced between the two sheared waveforms, to make the interpretation of the fringes easier.

8.3.2. Interpretation of shearing interferograms

The interpretation of a lateral shearing interferogram is more difficult than that of an interferogram obtained with a Twyman–Green interferometer,

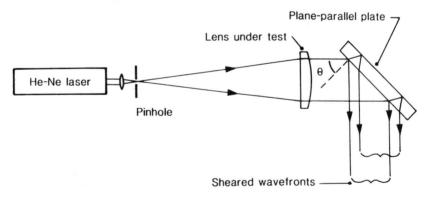

Fig. 8.5. Lateral shearing interferometer using a laser source and a tilted, plane-parallel plate [Murty, 1964].

since interference takes place between two aberrated wavefronts instead of between the aberrated wavefront and a perfect reference wavefront. The analysis of such an interferogram has been discussed by Malacara and Mendez [1968] and by Rimmer and Wyant [1975]. A polynominal $\Delta W(x,y)$ is fitted to the measured values of the optical path difference in two interferograms with mutually perpendicular directions of shear; the coefficients of $W(x,y)$ the polynominal representing the errors of the test wavefront, can then be derived from the coefficients of $\Delta W(x,y)$.

8.3.3. Rotational and radial shearing interferometers

As shown in Fig. 3.5, other forms of shear besides a lateral shear are possible. One special case is rotational shear in which interference takes place between two images of the test wavefront, one of which is rotated with respect to the other [Armitage and Lohmann, 1965; Murty and Hagerott, 1966]. Another is reversal shear [Gates, 1955; Saunders, 1955]. Perhaps the most useful is radial shear in which one of the images of the wavefront is contracted or expanded with respect to the other. [Brown, 1959; Hariharan and Sen, 1961e].

A number of optical arrangements for radial shearing interferometers are available; most of these have been described by Malacara [1978b]. With a thermal light source, a convenient arrangement is based on the use of a triangular path interferometer in which two images of the test wavefront of different sizes are produced by a lens system which is traversed in opposite directions by the two beams [Hariharan and Sen, 1961e]. With a laser source, a much simpler system can be used, consisting essentially of a thick lens [Steel, 1975]. In this, as shown in Fig. 8.6, interference takes place between the directly transmitted wavefront and the wavefront which has undergone one reflection at each surface. An alternative arrangement in which interference takes place between the wavefronts reflected from the front and back surfaces has been described by Zhou [1984].

To analyse a radial shearing interferogram it is convenient to express the deviations of the wavefront under test from a reference sphere in the form

$$W(\rho,\theta) = \sum_{n=0}^{k} \sum_{l=0}^{n} \rho^n \, (a_{nl} \cos^l \theta + b_{nl} \sin^l \theta) \qquad (8.3)$$

where ρ and θ are polar coordinates over a circle of unit radius defining the pupil. If the wavefront with which it is compared is magnified by a factor $(1/S)$, where S (<1) is known as the shear ratio, the deviations of this wavefront from the same reference sphere are

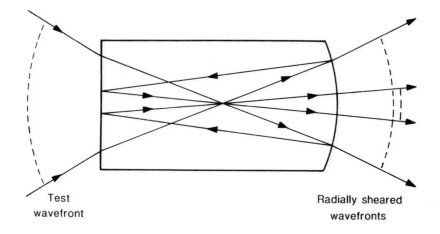

Fig. 8.6. Radial-shear interferometer for use with a laser source [Steel, 1975].

$$W' (\rho,\theta) = \sum_{n=0}^{k} \sum_{l=0}^{n} S^n \rho^n (a_{nl} \cos^l \theta + b_{nl} \sin^l \theta) \qquad (8.4)$$

Accordingly, the optical path differences in the interferogram are given by the relation

$$\begin{aligned}p(\rho,\theta) &= W(\rho,\theta) - W' (\rho,\theta) \\ &= \sum_{n=0}^{k} \sum_{l=0}^{n} (1-S^n) \rho^n (a_{nl} \cos^l \theta + b_{nl} \sin^l \theta)\end{aligned} \qquad (8.5)$$

It is apparent from a comparison of Eq. (8.5) with Eq. (8.3) that, with a reasonably small value of the shear ratio ($S<0.5$), the radial shear interferogram is very similar to the interferogram that would be obtained with a Twyman–Green interferometer, and the wavefront aberrations can be computed in a very similar fashion [Hariharan and Sen, 1961e; Malacara, 1974].

8.4. Grating interferometers

Gratings can be used instead of thin-film beam-splitters in any of the conventional two-beam interferometers such as the Michelson, Mach–Zehnder or Sagnac interferometers. Such an arrangement is very stable, since the angle between the two beams is affected only to a small extent by the orientation of the grating. With reflecting gratings it is possible to build an interferometer for use in the infrared or far ultraviolet regions, where good quality thin-film beam-splitters are not available [Munnerlyn, 1969].

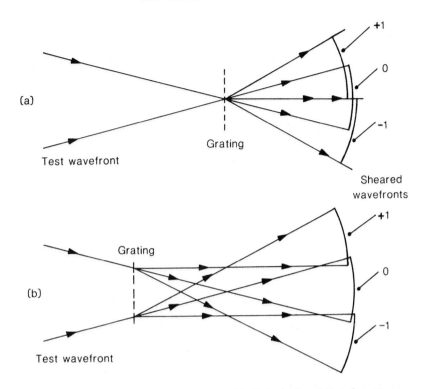

Fig. 8.7. Production of (a) shear, and (b) shear with tilt in the Ronchi interferometer.

A very simple type of grating interferometer is based on the Talbot effect. It can be shown that when a grating is illuminated with collimated monochromatic light, a series of images of the grating are formed at distances

$$z_m = 2m\Lambda^2/\lambda \qquad (8.6)$$

from it, where Λ is the period of the grating and m is an integer. This phenomenon is also known as Fourier- or self-imaging. If another identical grating is placed in the plane of one of these images and rotated slightly, high-contrast moiré fringes are obtained. If, then, a phase object is introduced between the gratings, close to the first grating, the moiré fringes are deformed and correspond to lines of equal deflection of the rays [Lohmann and Silva, 1971; Yokozeki and Suzuki, 1971]. The smallest detectable deflection is

$$\alpha_n = \Lambda/2z_m$$
$$= \lambda/m\Lambda \qquad (8.7)$$

The sensitivity, therefore, increases with the separation of the gratings.

Similar interference phenomena can also be seen even with an extended white light source (the Lau effect). In this case, coloured moiré fringes are formed. The theory of these fringes has been discussed by Jahns and Lohmann [1979], who have shown their similarity to Talbot fringes, and by Sudol and Thompson [1979]. They can also be used to study phase objects with a modified interferometric system in which one grating is imaged on the other by a telecentric optical system [Bartelt and Jahns, 1979].

Another simple type of grating interferometer is the Ronchi interferometer [see the review by Ronchi, 1964]. As shown in Fig. 8.7(a), a coarse grating (about 10 lines/mm) is placed at the focus of a converging wavefront from the system under test, and the resulting interference fringes are observed from behind the grating. This arrangement can be regarded as a shearing interferometer in which interference takes place between the directly transmitted wave and the diffracted orders which overlap it. The shear is determined by the period of the grating, while the tilt introduced between the interfering wavefronts can be increased by moving the grating away from the focus, as shown in Fig. 8.7(b).

The Ronchi interferometer is extremely easy to set up, but has the drawback that, for small shears, several different orders overlap and interfere with each other. One solution to this problem is the use of a double-frequency grating [Wyant, 1973]. The lower of the two grating frequencies is chosen so that the diffracted orders from both the gratings are separated from the directly transmitted beam, while the difference in the grating frequencies is chosen to give the required shear. A better solution is, as shown in Fig. 8.8(a), to use two gratings with the same spacing [Hariharan et al., 1974; Rimmer and Wyant, 1975]. In this case, as shown in Fig. 8.8(b), the shear can be varied by rotating one grating in its own plane. In addition, the amount and direction of the tilt introduced can be controlled by varying the separation of the gratings and their position with respect to the focus. A modified form of the two-grating interferometer which can be used with collimated beams has been described by Hariharan and Hegedus [1974].

8.5. The scatter-plate interferometer

The scatter-plate interferometer of Burch [1953, 1972] makes use of diffraction, but gives a fringe pattern which can be interpreted directly. A detailed description of its use has been given by Scott [1969]. To understand its operation, consider the optical system shown in Fig. 8.9, which can be used to test a concave mirror. In this arrangement, a lens L_1 forms an image of a point source on the mirror under test through a scatter plate S_1. This is

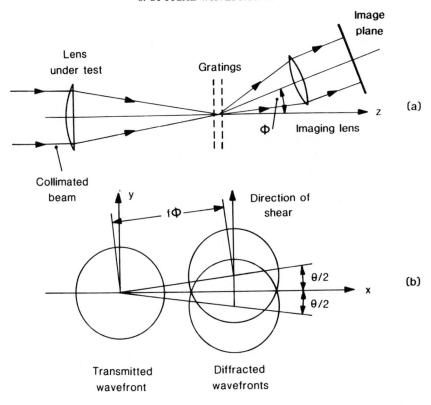

Fig. 8.8. Double-grating interferometer: (a) optical system and (b) sheared wavefronts [Hariharan *et al*, 1974].

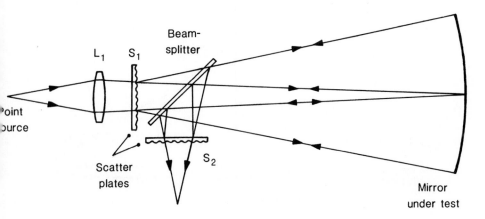

Fig. 8.9. Scatter-plate interferometer used to test a concave mirror.

a weak diffuser which transmits part of the light and scatters the rest to fill the aperture of the mirror. The scattered light is brought back to a focus by means of a small beam-splitter at another scatter plate S_2, which is identical to S_1 and is positioned so that it coincides with its image. Interference takes place between the wavefront transmitted by S_1 and scattered by S_2 and the wavefront scattered by S_1 and transmitted by S_2. The interference pattern seen is similar to that produced in a Twyman–Green interferometer, since the wavefront from the mirror interferes with a virtually perfect reference wavefront reflected from its central zone. A displacement of S_1 or S_2 along the axis changes the radius of curvature of the reference wavefront, while tilt can be introduced between the wavefronts by moving S_1 or S_2 in its own plane.

A very similar arrangement is possible using Fresnel zone plates [Murty, 1963; Smartt, 1974], which behave like lenses with multiple foci.

8.6. The point-diffraction interferometer

Another very simple common-path interferometer is the point-diffraction interferometer [Smartt and Steel, 1975]. In this interferometer, as shown in Fig. 8.10, a pinhole in a partially transmitting film is placed at the focus of a converging wavefront from the system under test. Interference takes place between the test wavefront, which is transmitted through the film, and a reference wave produced by diffraction at the pinhole. With a film having a

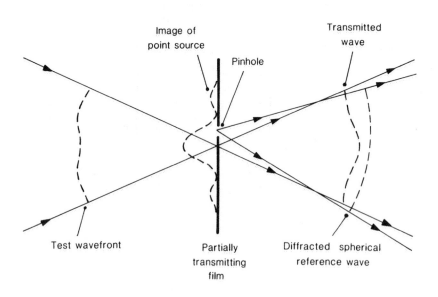

Fig. 8.10. Point-diffraction interferometer [Smartt and Steel, 1975].

transmittance of about 0.01, the amplitudes of the two wavefronts are very nearly equal, and fringes of good contrast can be obtained. If the pinhole is smaller than the Airy disc for an aberration-free wavefront, the reference wave is a spherical wave and the fringe pattern resembles a Twyman–Green interferogram. A major advantage of this interferometer is that it can be used to test a telescope objective under actual working conditions, with light from a bright star.

8.7. Polarization interferometers

Polarization interferometers use birefringent elements as beam-splitters [see Françon and Mallick, 1971]. They can be classified under three broad categories: systems that produce lateral shear with plane wavefronts, systems that produce lateral shear with spherical wavefronts and systems that produce radial shear.

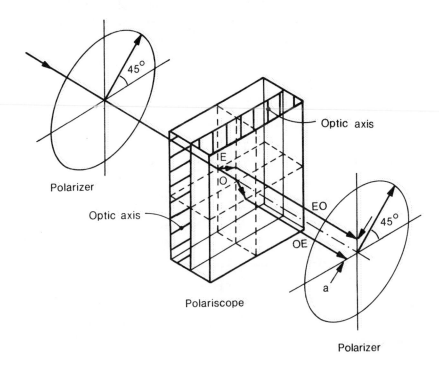

Fig. 8.11. The Savart polariscope.

8.7.1. Lateral shear with plane wavefronts

With a plane wavefront, lateral shear can be produced most conveniently with a Savart polariscope. This consists, as shown in Fig. 8.11, of two identical plates cut from a uniaxial crystal (usually calcite or quartz) with the optic axis at approximately 45° to the entrance and exit faces and put together with their principal sections crossed. A ray, linearly polarized at 45° by means of a polarizer P_1, incident on the first plate, is split into two rays, the ordinary (O) and the extraordinary (E). Because the second plate is rotated by 90° with respect to the first plate, the ordinary ray in the first plate becomes the extraordinary (OE) in the second, while the extraordinary ray in the first becomes the ordinary (EO) in the second. Both rays therefore emerge parallel to each other, but with a mutual displacement

$$a = 2^{1/2} \, d(n_e^2 - n_o^2)/(n_e^2 + n_o^2) \qquad (8.8)$$

where d is the thickness of the plates and n_o and n_e are the ordinary and extraordinary indices of refraction of the crystal. Since these rays are

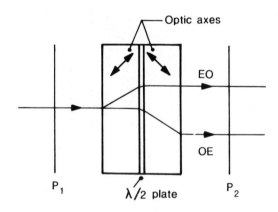

Fig. 8.12.   Savart polariscope modified to give a wider field [Françon, 1957].

orthogonally polarized, a second polarizer P_2 is used to make them interfere. For small angles of incidence, the interference pattern consists of straight, equally spaced fringes localized at infinity, with an angular separation $\alpha = \lambda/a$.

Straight fringes can be obtained over a much wider field with a modified Savart polariscope [Françon, 1957]. This consists, as shown in Fig. 8.12, of two identical plates cut, as before, at about 45° to the optic axis, but with one plate rotated through 180° so that, while the principal sections of the two plates are parallel, the optic axes are perpendicular. A half-wave plate is inserted between the two plates with its axis at 45° to the principal sections. In this case, the rays stay in the same plane and their separation is

$$a = 2d(n_e{}^2 - n_o{}^2)/(n_e{}^2 + n_o{}^2) \qquad (8.9)$$

A variable shear can be obtained with this polariscope if the polarizers P_1, P_2 are fixed to the plates and the two components are rotated in opposite directions through the same angle [Steel, 1964a]; if this angle is θ, the lateral shear is

$$a = 2d\left[(n_e{}^2 - n_o{}^2)/(n_e{}^2 + n_o{}^2)\right] \cos \theta \qquad (8.10)$$

Two identical Savart polariscopes of the type shown in Fig. 8.12 can be used in a system such as that shown in Fig. 8.13 for the examination of phase objects. A half-wave plate with its optic axis at 45° to the principal sections of the two polariscopes Q_1, Q_2 is interposed between them. As a result, the OE ray emerging from Q_1 becomes the EO ray in Q_2 and *vice versa*, so that the shear produced by Q_1 is compensated by Q_2. Because of this, it is possible to use an extended white-light source with this system.

8.7.2. Lateral shear with spherical wavefront

With a spherical wavefront, lateral shear can be produced by a Wollaston prism. This is made up, as shown in Fig. 8.14(a) of two prisms having the same angle θ cut from a uniaxial crystal in such a way that the optic axes are parallel to the outer faces but are at right angles to each other. An incident ray is split into two rays making an angle α with each other, which, when θ is small, is given by the relation,

$$\alpha = 2(n_e - n_o) \sin \theta \qquad (8.11)$$

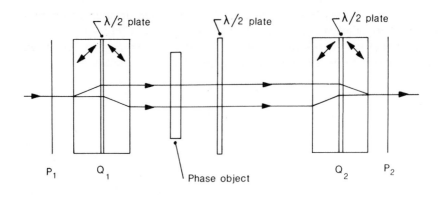

Fig. 8.13. Shearing interferometer using two Savart polariscopes.

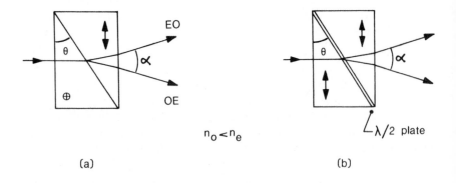

Fig. 8.14. (a) The Wollaston prism, and (b) modification to give a wider field.

A wider useful field can be obtained with the modified form of the Wollaston prism shown in Fig. 8.14(b). In this, the optic axis is parallel in the two elements, and a half-wave plate is inserted between them. Lateral shearing interferometers using a Wollaston prism have been described by Philbert [1958] and by Dyson [1963].

8.7.3. Radial shear

A lens made from a birefringent crystal produces two images at different points along the axis and can be used to obtain radial shear.

Radial shear interferometers using such birefringent lenses have been described by Dyson [1957] and by Steel [1965].

8.8. Interference microscopy

Two-beam interference microscopes are widely used in the study of transparent living objects which cannot be stained without damaging them.

Early instruments were based on the Michelson and Mach–Zehnder interferometers. However, these have now been replaced largely by common-path interference microscopes with double-focus or lateral-shearing systems using a Sagnac interferometer or birefringent elements. Details of several of these instruments are to be found in works by Françon [1961], Françon and Mallick [1971] and Krug, Rienitz and Schulz [1964] as well as in a review by Bryngdahl [1965]. With a relatively large shear, it is possible to have one beam passing through a clear area on the slide, so that an interference pattern corresponding to the actual variations of the optical thickness of the specimen is obtained.

Alternatively, the shear may be made very small to give the gradient of the optical thickness. A very simple interference microscope based on the point-diffraction interferometer has been described by Ross and Singh [1983].

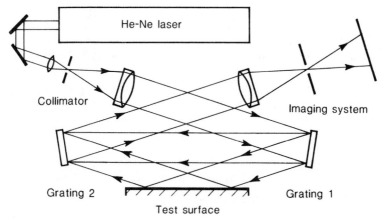

Fig. 8.15. Grazing-incidence interferometer using reflection gratings [Hariharan, 1975].

8.9. Testing of non-optical surfaces

The natural contour interval of half a wavelength (about 0.25 μm with visible light) obtained with conventional interferometers makes it difficult to apply them to measurements on surfaces whose deviations from flatness, or from a reference sphere, exceed about 10 μm. It is also not possible to obtain fringes with surfaces which are rough on an optical scale.

One method of solving these problems is to use a longer wavelength. Infrared interferometry with a CO_2 laser at a wavelength of 10.6 μm has been used successfully to test ground surfaces as well as surfaces with relatively large deviations from a reference sphere [Munnerlyn and Latta, 1968; Kwon et al., 1980].

A simpler alternative with nominally flat surfaces is to reduce the sensitivity of the interferometer by using collimated light incident obliquely, instead of normally, on the surface [Linnik, 1942]. The contour interval is then $\lambda/2 \cos \theta$, where θ is the angle of incidence, and fringes can be obtained with surfaces such as fine-ground glass and metal [Abramson, 1969a; Birch 1973]. The low reflectivity of the test surface can be compensated by means of a system such as that shown in Fig. 8.15 using blazed reflection gratings to divide and recombine the beams [Hariharan, 1975a], or by using a pair of wedged beam splitters, coated on only one side [Murty and Shukla, 1976].

8.10. Measurement of the optical transfer function

One way of evaluating the quality of an optical system is by measurements of its pupil function, the deviation of the wavefront at the exit pupil from a sphere centred on the Gaussian image point. Another method is by measurements of the intensity distribution in the image of a point. This intensity distribution is known as the point-spread function of the system. Yet another way of specifying the performance of an optical system is by the Fourier transform of its point-spread function, which is known as the optical transfer function (OTF). This is a measure of the transmittance of the system for different spatial frequencies in the object and has the advantage that, for a system built up of a number of linear elements, the OTF of the system is given by the product of the OTF's of the individual elements.

It can be shown (see Appendix A.2) that with coherent illumination $g(\xi,\eta)$, the pupil function of the system, and $G(x,y)$, the amplitude distribution in the image of an infinitely distant point, are related by a Fourier transform, so that

$$G(x,y) \leftrightarrow g(\xi,\eta) \qquad (8.12)$$

With incoherent illumination, the intensity distribution in the image is given by $|G(x,y)|^2$, so that the OTF is

$$\Omega(\xi,\eta) = \mathscr{F}\{|G(x,y)|^2\}$$
$$= g(\xi,\eta) \star g(\xi,\eta) \qquad (8.13)$$

where $\star$ denotes the correlation operator (see Appendix A.1). The right-hand side of Eq. (8.13) is usually normalized so that it has a maximum value of unity.

The OTF can be measured directly by a shearing interferometer [Hopkins, 1955]. To understand how this is done, consider two images of the wavefront leaving the pupil of the system under test, which have been laterally sheared by an amount η_s along (say) the η axis, and with an additional phase difference ϕ introduced between them. The complex amplitudes in these two wavefronts at a point (ξ,η) are $g(\xi,\eta)$ and $g(\xi,\eta - \eta_s)$ exp iϕ, respectively, and the shear η_s corresponds to a spatial frequency s defined by the relation $s = \eta_s/\lambda f$. The total flux leaving the interferometer, which is also the total flux in the image of a point source, is then

$$I(\phi) = \int\int |g(\xi,\eta) + g(\xi,\eta-\eta_s)\exp(i\phi)|^2 d\xi d\eta$$

$$= \int\int |g(\xi,\eta)|^2 d\xi d\eta + \int\int |g(\xi,\eta-\eta_s)|^2 d\xi d\eta$$

$$+ 2\,\text{Re}\,[\exp(i\phi)\int\int g(\xi,\eta)\,g^*(\xi,\eta-\eta_s)d\xi d\eta]$$

$$= 2\bar{I}[1 + |\Omega\,s)|\cos(\psi+\phi)] \tag{8.14}$$

where $\bar{I} = \int\int |g(\xi,\eta)|^2 d\xi d\eta = \int\int |g(\xi,\eta-\eta_s)|^2 d\xi d\eta$ and $\Omega(s) = |\Omega(s)|\exp(i\psi)$

Equation (8.14) shows that if the additional phase shift ϕ between the two sheared wavefronts is varied linearly, the transmitted flux varies sinusoidally. The difference between the maximum and minimum values of $I(\phi)$ divided by their average $\bar{I}$ gives the modulus $|\Omega(s)|$ of the OTF (known as the modulation transfer function, or the MTF) for the spatial frequency s, while its phase ψ can be evaluated from the values of ϕ corresponding to these maxima and minima. These measurements are repeated at increasing values of the shear to trace out the OTF curve.

To carry out these measurements we require an interferometer in which the shear can be set at any desired value without introducing tilt between the beams, while the path difference is varied over a convenient range. Early systems used modified Twyman–Green [Kelsall, 1959], Sagnac [Hariharan and Sen. 1960d], polarizing [Tsuruta, 1963] and Mach–Zehnder [Montgomery, 1964] interferometers. Perhaps the simplest system is that due to Wyant [1976] which is shown in Fig. 8.16. In this arrangement, the beam transmitted through a tilted plane-parallel plate and the beam reflected off its two surfaces give rise to a lateral shear interferogram. The shear is then

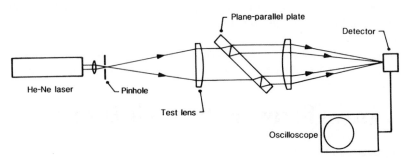

Fig. 8.16. Parallel-plate shearing interferometer for measurements of the MTF of a lens [Wyant, 1976].

$$s = (n^2 - \sin^2\theta)^{-\frac{1}{2}}d \sin 2\theta \qquad (8.15)$$

while the optical path difference is

$$p = 2nd\{[(n^2 - \sin^2\theta)/n^2]^{\frac{1}{2}} - 1\} \qquad (8.16)$$

where d is the thickness of the plate and θ is the angle of incidence. The two beams are brought to a focus on a detector, whose output is displayed on an oscilloscope, as a function of θ, as the plate is rotated. The envelope of the time-varying signal on the oscilloscope then gives a direct display of the MTF of the system under test.

9
Interferometry with lasers

Lasers give light of much greater coherence and intensity than thermal sources. As we have seen in Chapter 7 and Chapter 8, this has removed many early limitations in length interferometry, as well as in the application of interferometers to optical testing. In addition, it has led to the development of several new and interesting techniques, some of which will be discussed in this chapter.

9.1. Photoelectric setting methods

Photoelectric detection has been used for many years to obtain increased precision in length measurements. The most common technique is a null method based on introducing a small modulation of the optical path difference [Baird, 1954; Bruce and Hill, 1961]. To implement it, the fringe pattern is imaged on a small aperture, and the transmitted flux is analysed into its harmonic components. This disappearance of the component at the modulation frequency then defines a setting on an intensity maximum or minimum. The precision attainable with this technique has been discussed by Hill and Bruce [1962, 1963], Hanes [1963], and Ciddor [1973], who have shown that with light from a thermal source the major limitations arise from its limited coherence and low intensity. The use of a laser gives fringes of high visibility, with a limiting precision, in an ideal case, of the order of 1 part in 10^{12}. The precision of such measurements is therefore limited, in practice, largely by the mechanical stability of the interferometer.

9.2. Fringe counting

Another technique, which was first described several years ago by Peck and Obetz [1953], but has now become increasingly important, is electronic fringe counting. Basically, this involves an optical system giving two uniform interference fields in one of which there is an additional phase difference of $\pi/2$ between the interfering beams. These can be produced conveniently by using a beam-splitter with a multilayer semireflecting metal coating [Raine and Downs, 1978], in which case the interference pattern formed by the normal output beams and that formed by the beams going back to the source meet this condition. Two detectors viewing these two fields provide two signals in quadrature which can drive a bidirectional counter. The same signals can also be used to determine the fractional interference order. One way of doing this is to take these signals to the horizontal and vertical deflection amplifiers of an oscilloscope. A circular pattern is obtained, which is traversed once each time the phase difference between the beams changes by 2π. A fringe counting interferometer of this type using a He–Ne laser has been described by Gilliland et al. [1966].

An alternative method of fringe counting is based on polarization coding [Dyson et al., 1970]. In this technique, the two beams emerging from the interferometer, are linearly polarized at right angles to each other and traverse a $\lambda/4$ plate oriented at 45°, which converts one beam into right-handed and the other into left-handed circularly polarized light. If the amplitudes of the two beams are the same, they add to produce a linearly polarized beam whose plane of polarization rotates through 360° for a change in the optical path difference between the beams of two wavelengths. The rotation of the plane of polarization can be followed by a polarizer controlled by a servo system [Roberts, 1975]. The effect of dc drifts can be minimized by modulating the output intensity by a Pockels cell or a Faraday cell and using a phase sensitive detector. A detailed analysis of the possible errors in such a system (which is also known as an 'optical screw') has been made by Hopkinson [1978].

A drawback of both these fringe counting systems is that the inherent low-frequency noise in a laser can affect their operation. This problem can be avoided with a system that uses two different optical frequencies [Dahlquist et al., 1966; Dukes and Gordon, 1970]. The two optical frequencies are generated by a single He–Ne laser, which is forced to oscillate simultaneously on two frequencies, ν_1 and ν_2, separated by a constant difference of about 2 MHz, by the application of an axial magnetic field. The beam from the laser consisting of these two waves, which are circularly polarized in opposite senses, goes through a $\lambda/4$ plate, which converts them to orthogonal linear polarizations, and a beam-expanding telescope before entering the interferometer.

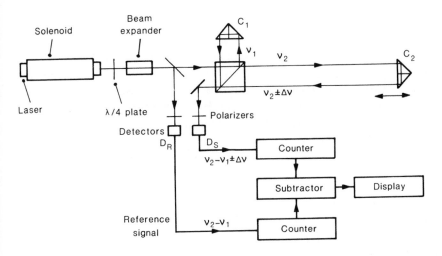

Fig. 9.1. Fringe-counting interferometer using a two-frequency laser [after Dukes and Gordon, 1970]. © Copyright 1970 Hewlett-Packard Company. Reproduced with permission.

In the interferometer, as shown schematically in Fig. 9.1, a portion of the beam is split off initially and, after going through a polarizer which mixes the two frequencies, is incident on a detector D_R. The output from this detector, which is at the beat frequency ($\nu_2 - \nu_1$), constitutes a reference frequency. The main beam goes to a polarizing beam-splitter at which one frequency (say ν_1) is reflected to a fixed cube corner C_1, while the other, ν_2, is transmitted to a movable cube corner C_2 which constitutes the measuring element. Both frequencies then return along a common axis and, after passing through a polarizer, are incident on another detector D_S.

When both the cube corners are stationary, the output signal from the detector D_S is also at the beat frequency ($\nu_2 - \nu_1$). However, if the cube corner C_2 is moved slowly, the frequency ν_2 is shifted up or down because of the Doppler effect. Typically, a velocity of 0.1 m/s causes a shift $\Delta\nu$ of approximately 300 kHz. As a result, the frequency at the output from the detector D_S becomes ($\nu_2 - \nu_1 \pm \Delta\nu$).

The outputs from the two photodetectors D_R and D_S go to a differential counter. One frequency drives the counter forward while the other drives it backwards. If the cube corner C_2 is stationary, the two frequencies are equal and no net count accumulates. However, if C_2 is moved, a net positive or negative count is produced, which gives the change in optical path in wavelengths of light.

This interferometer can be used for measurements over paths as long as 60 m. Its application to precision measurements in a workshop has been described by Liu and Klinger [1979].

9.3. Heterodyne interferometry

In this technique [Crane, 1969], a small frequency shift is introduced in one beam of the interferometer by means of a rotating $\lambda/4$ plate in series with a fixed $\lambda/4$ plate. Other systems which have been used for the same purpose include a rotating grating and two acousto-optic modulators operated at slightly different frequencies. The electric fields corresponding to the two beams at any point $P(x,y)$ in the field can be written as

$$E_1(x,y,t) = |a_1(x,y)| \exp \{i[2\pi\nu_1 t - \phi_1 (x,y)]\} \tag{9.1}$$

and

$$E_2(x,y,t) = |a_2(x,y)| \exp \{i[2\pi\nu_2 t - \phi_2(x,y)]\} \tag{9.2}$$

where ν_1, ν_2 are the two optical frequencies, $|a_1 (x,y)|$, $|a_2 (x,y)|$ are the real parts of the amplitudes of the two waves, and $\phi_1 (x,y)$, $\phi_2 (x,y)$ are their phases.

The resultant intensity at P is then

$$\begin{aligned} I (x,y,t) &= |E_1(x,y,t) + E_2(x,y,t)|^2 \\ &= |a_1(x,y)|^2 + |a_2(x,y)|^2 \\ &\quad + 2|a_1(x,y)||a_2(x,y)| \cos [2\pi(\nu_1 - \nu_2)t - \Delta\phi] \end{aligned} \tag{9.3}$$

where $\Delta\phi = \phi_2(x,y) - \phi_1(x,y)$. It follows from Eq. (9.3) that the output from a detector at this point will be modulated at a frequency $(\nu_1 - \nu_2)$. The phase of this modulation corresponds to the original phase difference between the interfering wavefronts. It can be measured electronically with respect to a reference signal derived either from a second detector to which the optical paths remain unchanged, or from the source driving the modulator. This technique is capable of very high precision and has been used for a variety of measurements [Ohtsuka and Sasaki, 1974; Lavan et al., 1976; Kristal and Peterson, 1976].

The heterodyne method can also be used to measure small changes in the separation of the mirrors of a Fabry–Perot interferometer. In this technique, the frequency of a slave laser is locked to a transmission peak of the interferometer so that the laser wavelength is an integral submultiple of the optical path difference in the interferometer. Any variation in the separation of the mirrors then results in a corresponding variation in the laser wavelength and, hence, in its frequency. To detect these variations the beam from the slave laser is mixed at a fast photodiode with the beam from a reference laser whose frequency has been stabilized, and the beat frequency is measured. This technique has been used, as described in Section 7.10, in a repetition of the Michelson–Morley experiment.

As can be seen, the heterodyne technique replaces measurements of lengths by measurements of frequency. Since measurements can be made in the frequency domain with very high precision, this is a very powerful approach.

9.4. Digital interferometry

For many years, the chief disadvantage of interferometry as a tool for optical testing has been the labour involved in analysing an interferogram. The development of digital computers has completely changed this situation, and interferometers using digital techniques to process video images are now being used more and more widely for high-precision tests of optical elements.

Electronic measurement of fringe patterns is possible either by plotting the loci of the centres of the fringes [Womack et al., 1979], using a television camera linked to a digital computer, or by directly measuring the phase difference between the two interfering wavefronts at an array of points covering the field. The latter approach is preferable, since it is capable of higher precision and is not affected by variations in the illumination over the field. The techniques described for this purpose fall into two broad groups.

One approach is based on point-by-point measurements using mechanical or optical scanning of the fringe pattern. Measurements of the phase difference between the interfering wavefronts can be carried out either by a null-detection technique using a small phase modulation of one of the beams [Pearson et al., 1976; Johnson et al., 1977], or by heterodyne techniques [Sommargren and Thompson, 1973; Massie et al., 1976]. The time involved in making such measurements can be reduced by using an image dissector camera to scan the interference pattern [Mottier, 1979; Massie, 1980].

Another approach is based on measurements on a number of points in parallel, using a charge-coupled-detector (CCD) array. This approach has the advantage that a much better signal-to-noise (S/N) ratio is possible at low light levels, since the signal at each element is integrated over the whole measurement period. Figure 9.2 is a schematic of a typical system which has been used for this purpose.

In one method [Wyant, 1975; Stumpf, 1979; Schaham, 1981], the optical path difference between the interfering wavefronts is made to vary linearly with time by moving one of the end mirrors of the interferometer, which is mounted on a piezoelectric translator (PZT). The intensity at any point $P(x,y)$ in the interference pattern can then be written as a function of time

$$I(x,y) = I_0(x,y) \{ 1 + \mathcal{V} \sin [(2\pi t/T) + \phi(x,y)]\} \tag{9.4}$$

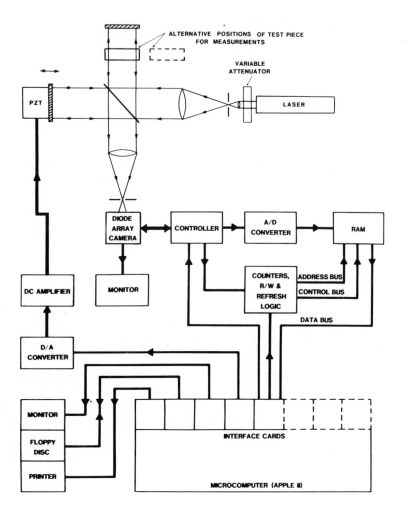

Fig. 9.2. Schematic of a typical system for digital interferometry [Hariharan *et al.*, 1984].

where $I_0(x,y)$ is the average intensity at P, $\mathscr{V}$ is the visibility of the fringes, T is the period of the resulting modulation and the phase difference between the two interfering wavefronts at P when t = 0 is $\phi(x,y) - \pi/2$.

To determine $\phi(x,y)$, the integrated output from each of the elements is read out at the end of four equal intervals covering one such period. If we neglect a constant of proportionality, these outputs can be written as

$$A = E[(1/4) + (1/2^{1/2} \pi) \, \mathscr{V} \sin \phi(x,y)] \qquad (9.5)$$

$$B = E[(1/4) + (1/2^{1/2}\pi) \mathscr{V} \cos \phi(x,y)] \tag{9.6}$$

$$C = E[(1/4) - (1/2^{1/2}\pi) \mathscr{V} \sin \phi(x,y)] \tag{9.7}$$

and

$$D = E[(1/4) - (1/2^{1/2}\pi) \mathscr{V} \cos \phi(x,y)] \tag{9.8}$$

where

$$E = \int_{-T/8}^{T/8} I(x,y)\mathrm{d}t + \int_{T/8}^{3T/8} I(x,y)\mathrm{d}t + \int_{3T/8}^{5T/8} I(x,y)\mathrm{d}t + \int_{5T/8}^{7T/8} I(x,y)\mathrm{d}t$$

$$= A + B + C + D \tag{9.9}$$

Accordingly, it follows that

$$\tan \phi(x,y) = (A - C)/(B - D) \tag{9.10}$$

This system requires the variation of the optical path with time to be strictly linear, though the effect of nonlinearities can be minimized by increasing the number of integration periods per cycle to eight.

In another method used by Bruning et al. [1974], the optical path difference is changed in equal steps by means of the mirror mounted on the PZT, and the corresponding values of the intensity at each point in the interference pattern are recorded. If the complex amplitudes of the reference and test wavefronts are written as

$$a(x,y) = a_0 \exp(-ikp) \tag{9.11}$$

and

$$b(x,y) = b_0 \exp[-ik\, W(x,y)] \tag{9.12}$$

where $W(x,y)$ represents the profile of the test wavefront, the intensity in the interference pattern is

$$I(x,y,p) = a_0^2 + b_0^2 + 2a_0b_0 \cos k[W(x,y) - p] \tag{9.13}$$

An alternative representation for Eq. (9.13) is

$$I(x,y,p) = f + g_1 \cos k\ell + h_1 \sin k\ell \tag{9.14}$$

where the coefficients f, g_1 and h_1 are functions of x and y. This is a Fourier series consisting only of the dc term and the first harmonics. Accordingly, the coefficients at any point can be found by measuring the intensity $I(x,y,p)$ for values of p given by

$$p = p_j = j\lambda/2m \qquad (9.15)$$

where $j = 1,2,\ldots mq$, and m and q are integers. We then have

$$\begin{aligned} g_1 &= (2/mq) \sum_{j=1}^{mq} I(x,y,p_j) \cos kp_j \\ &= 2a_0b_0 \cos k\,W(x,y) \end{aligned} \qquad (9.16)$$

and

$$\begin{aligned} h_1 &= (2/mq) \sum_{j=1}^{mq} I(x,y,p_j) \sin kp_j \\ &= 2a_0b_0 \sin k\,W(x,y) \end{aligned} \qquad (9.17)$$

so that

$$W(x,y) = (1/k) \tan^{-1} (h_1/g_1) \qquad (9.18)$$

While as many as 100 readings ($m = 25, q = 4$) have been used, a typical implementation of this technique uses an eight-step staircase modulation ($m = 8, q = 1$) of the optical path difference, and can give values of $W(x.y)$ over a 32 × 32 array of points to better than $\lambda/100$.

A simpler method [Carré, 1966] involves only four measurements of the intensity at a point corresponding to four values of an additional phase shift. If these four values are equally spaced, the original phase difference between the beams as well as the magnitude of the phase steps can be calculated. A further simplification is possible if the phase steps are known, in which case only three measurements of the intensity are necessary. Typically, one value of the phase step can be zero, while the other two are $\pi/2$ and π [Frantz et al., 1979; Dörband, 1982]. In this case, we have

$$I(0) = |a_1|^2 + |a_2|^2 + 2|a_1||a_2| \cos \phi \qquad (9.19)$$

$$I(\pi/2) = |a_1|^2 + |a_2|^2 + 2|a_1||a_2| \sin \phi \qquad (9.20)$$

and

$$I(\pi) = |a_1|^2 + |a_2|^2 - 2|a_1||a_2| \cos \phi \qquad (9.21)$$

so that

$$\tan \phi = [2I(\pi/2) - I(0) - I(\pi)]/[I(\pi) - I(0)] \qquad (9.22)$$

Alternatively, it is possible to use a phase step of $\pm 2\pi/3$ [Hariharan et al., 1982]. In this case,

$$3^{-\frac{1}{2}}\tan \phi = [I(-2\pi/3) - I(2\pi/3)]/[2I(0) - I(2\pi/3) - I(-2\pi/3)] \qquad (9.23)$$

Because of the reduction in memory requirements as well as the simplicity of the algorithm for calculating the phase, a relatively inexpensive microcomputer can be used with such a system. Measurements covering a 100×100 array of points with a precision of $2\pi/200$ can be made in about 5 seconds.

Because measurements can be made so rapidly with such an interferometer, and the data can be stored, the effects of air currents and mechanical instability can be reduced by averaging a number of observations. Similarly, it is possible to eliminate systematic errors due to imperfections in the interferometer by subtracting, as shown schematically in Fig. 9.2, from the readings with the test piece, a set of readings made without it or with a standard test piece [Bruning, 1978; Hariharan et al., 1984]. A more serious problem, when making measurements to $\lambda/100$ or better, is scattered light and stray reflections within the interferometer. Because of the high spatial and temporal coherence of light from a single-frequency laser, the two beams in the interferometer can also interfere with weaker beams reflected from the surfaces of the various components. Unless care is taken to eliminate these unwanted reflections, they can give rise to distortions in the wavefront which is being examined, resulting in significant systematic errors.

9.5. Fibre interferometers

The use of single-mode optical fibres to build analogues of conventional two-beam interferometers became practical with the development of lasers. Since such a fibre does not allow higher order modes to propagate, a smooth wavefront is obtained at the output end. Fibre interferometers can give very high sensitivity for some types of measurements because they make possible very long paths in a small space and have very low noise, permitting the use of sophisticated detection techniques.

The first application of fibre interferometers was in rotation sensing, where, as described in Section 7.11, a closed multi-turn loop made of a

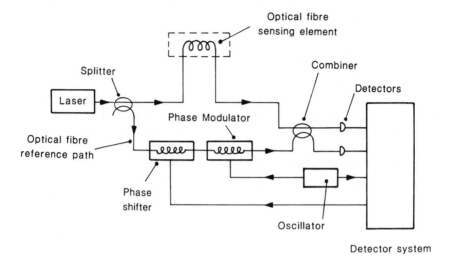

Fig. 9.3. Schematic of an interferometer using a single-mode fibre as a sensing element [Giallorenzi *et al.*, 1982]. © Copyright 1982, IEEE; reproduced with permission.

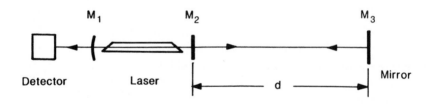

Fig. 9.4. Laser-feedback interferometer

single-mode fibre was used as an analogue of the Sagnac interferometer. A detailed analysis of the sensitivity possible with such rotation sensors has been made by Lin and Giallorenzi [1979], while Bergh *et al.* [1981] have described a system based entirely on fibre-optic components.

Other applications take advantage of the fact that the optical path length changes when a fibre is stretched or when the ambient pressure and temperature change, so that an optical fibre can be used as a sensor. Fibre interferometers which are analogues of the Michelson and Mach–Zehnder interferometers have been used to measure mechanical strains and pressure variations as well as temperature differences [Bucaro *et al.*, 1977; Butter and Hocker, 1978; Hocker, 1979 and Jackson *et al.*, 1980a]. Magnetic fields can also be measured either by using a magnetostrictive jacket on the fibre, or by bonding the fibre to a magnetostrictive element [Yariv and Winsor, 1980; Rashleigh, 1981]. Much of the work in these fields has been reviewed by Giallorenzi et al. [1982] and by Kyuma et al. [1982].

Figure 9.3 is a schematic of a generalized fibre interferometer showing the principal components of such a system. While early fibre interferometers used gas lasers, solid-state GaAlAs lasers are now employed widely because, apart from their small size and high output, they operate at a wavelength (λ ≈ 0.9 μm) at which the losses in silica fibres are much lower than in the visible region. The use of an optical layout which is analogous to the Mach–Zehnder interferometer is also well established, since this avoids light being reflected back to the laser, causing it to jump from one longitudinal mode to another. Normal beam-splitters have also been replaced by twisted-fibre optical couplers [Sheem and Giallorenzi, 1979] permitting an all-fibre arrangement with a considerable reduction in noise. Measurements are usually carried out either by a phase-tracking system or by a heterodyne system. Phase modulation is introduced by a fibre stretcher [Jackson et al., 1980b] in the reference beam, after which the two beams are combined, and the resulting signal is picked up by a detector followed by a suitable demodulator. Optical phase shifts of 10^{-6} radian can be detected with such a system [Dandridge and Tveten, 1981].

9.6. Laser-feedback interferometers

The laser-feedback or three-mirror interferometer [Ashby and Jephcott, 1963; Clunie and Rock, 1964] is a very simple device. It makes use of the fact that if, as shown in Fig. 9.4, a mirror M_3 is used to reflect the output beam back into the laser cavity, the laser output varies cyclically with d, the distance of M_3 from M_2, the laser mirror nearest to it.

The operation of such an interferometer can be analysed very roughly [Peek et al., 1967] by considering M_3 and M_2 as a Fabry–Perot interferometer, which effectively replaces M_2 as the end mirror of the laser cavity. The reflectance of this Fabry–Perot interferometer is

$$R = R_2 + (1 - R_2)\{1 - (1 - R_3)/[1 + R_2 R_3 + 2(R_2 R_3)^{1/2} \cos \phi]\} \quad (9.24)$$

where R_2 and R_3 are the reflectances of M_2 and M_3, and $\phi = 4\pi nd/\lambda$. It follows from Eq. (9.24) that the value of R is a maximum when $nd = m\lambda/2$, where m is an integer, and drops to a minimum when $nd = (2m + 1)\lambda/4$. Typically, if $R_2 = 0.99$ and $R_3 = 0.75$, $R_{max} = 0.999$ while $R_{min} = 0.87$. In addition, if ϕ is not equal to an integral multiple of π, there is a change in the effective length of the cavity.

If the laser is oscillating on a single transition, the resultant change in the gain merely enhances or diminishes the laser power on this transition. If, however, the laser is oscillating simultaneously on two transitions which share a common upper or lower level, and if M_3 reflects only one of these wavelengths, the output power at the other wavelength will vary in antiphase to the power at the first wavelength. This effect can be observed with a He-Ne laser and has been used to measure plasma densities at 3.39 μm with a detector operating at 633 nm [Ashby and Jephcott, 1963].

The main advantages of this type of interferometer are very easy alignment, inherent mechanical stability and high sensitivity, since a small change in the gain results in a large change in the output. A difficulty is that when the phase change exceeds π it is difficult to determine its sign. One solution is to use a laser oscillating on two longitudinal modes which are orthogonally polarized. If the ratio of the lengths of the external cavity and the laser cavity are properly chosen, these signals can be separated and used to obtain phase data in quadrature over a limited displacement range [Timmermans et al., 1978].

9.7. Diode-laser interferometers

A very compact laser-feedback interferometer can be set up with a single-mode GaAlAs laser and an external mirror whose position is to be monitored [Dandridge et al., 1980]. In one mode of operation, the laser current is held constant and the output power is monitored by a photodiode; alternatively, the drive current to the laser to maintain a constant output power is measured. Additional flexibility in applications is possible by using a single-mode fibre to couple the external mirror to the laser.

Since the response of such a system is not linear, its useful range is limited. Greater dynamic range as well as higher sensitivity can be obtained by mounting the mirror on a PZT and using an active feedback loop to hold the optical path from the laser to the mirror constant, at a suitable fixed point on the linear section of the power versus distance curve. Vibration amplitudes of the order of 10^{-4} nm can be measured with such an interferometer.

9.8. Interferometry with pulsed lasers

As described in Section 6.4, a Q-switched ruby laser or Nd:YAG laser can be used to produce light pulses with a duration less than 0.1 μs. Such a laser

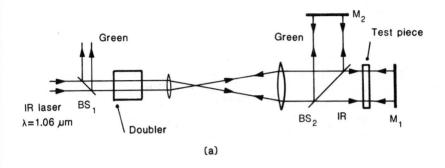

(a)

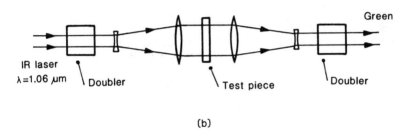

(b)

Fig. 9.5. Second-harmonic interferometers: (a) nonlinear Twyman–Green, (b) nonlinear Jamin [after Hopf, 1980].

can be used very effectively for interferometric studies of high-speed gas flows. A number of different types of interferometers for such studies have been reviewed by Tanner [1966, 1967], who has shown that the coherence requirements for most of these can be met quite easily by a commercial ruby laser.

The use of a pulsed nitrogen laser ($\lambda = 337$ nm), producing pulses with a duration less than 500 ps, for interferometric studies of short-lived plasmas has also been described by Schmidt et al. [1975]. Even shorter pulses can be produced by a number of techniques.

9.9. Second-harmonic interferometry

Another interesting technique which has been made feasible by the very high light intensity produced by a pulsed laser is second-harmonic interferometry [Hopf et al., 1980].

An interferometer of this type, which may be called a nonlinear Twyman–Green interferometer, is shown schematically in Fig. 9.5(a). This interferometer can be used to test an optical element made of a material, such as silicon, which transmits only in the near infrared, while producing an interferogram with visible light. In this instrument, the beam from a Q-switched Nd:YAG laser (λ = 1.06 μm) is incident on a frequency-doubling crystal. The green (λ = 0.53 μm) and infrared beams emerging from this crystal are then expanded by a pair of lenses and separated by a dichroic beamsplitter (BS_2), so that each beam traverses one arm of the interferometer. The test piece is placed in the infrared beam while the green beam serves as the reference. When the two beams return to the crystal, the green beam passes through it unchanged, while the infrared beam undergoes frequency doubling to generate a second green beam. Another dichroic beam-splitter (BS_1) then reflects the two green beams to the observation plane.

In this interferometer, the additional optical path difference between the beams, at any point $P(x,y)$ in the field, due to the test piece is

$$p(x,y) = 2(n_{IR} - 1)\, d(x,y) \qquad (9.25)$$

where $d(x,y)$ is the thickness of the test piece at P and n_{IR} is its refractive index at 1.06 μm. The resulting change in the interference order is, therefore,

$$\begin{aligned} N(x,y) &= p(x,y)/\lambda_G \\ &= 2(n_{IR} - 1)\, d(x,y)/\lambda_G \end{aligned} \qquad (9.26)$$

Since the same small fraction of the power in the infrared beam is doubled at each pass, the two green beams have the same intensity and fringes with good visibility are obtained.

Another interferometer of this type, which may be called a nonlinear Fizeau interferometer, is obtained if the beam-splitter BS_2 and the mirror M_2 in Fig. 9.5(a) are removed. In this case, since both the infrared and the green beams go through the object, there is, at any point $P(x,y)$ in the field, a difference in their optical paths

$$p(x,y) = 2(n_G - n_{IR})\, d(x,y) \qquad (9.27)$$

where n_G is the refractive index of the test piece at 0.53 μm. As before, when the two beams return to the crystal, the infrared beam undergoes frequency doubling to produce a second green beam which interferes with the first one. Dark fringes are obtained in the interference pattern when the optical path difference is

$$p(x,y) = (2m + 1) \lambda_G / 2 \qquad (9.28)$$

that is to say, when

$$d(x,y) = (2m + 1) \lambda_G / 4 (n_G - n_{IR}) \qquad (9.29)$$

An alternative form of this interferometer using two frequency-doubling crystals is shown in Fig. 9.5(b). This setup can be thought of as an analogue of the Jamin interferometer, since the two beams are separated and recombined in frequency space at the two crystals. An interferometer of this type could find use in applications such as the study of plasmas, where the dispersion is proportional to the electron density.

10
Interference spectroscopy

The instrumental function of a spectroscope is defined by its response $a(v-v')$ to a perfectly monochromatic spectral line $\delta(v-v')$. When the spectroscope is illuminated by a spectral line with a spectral energy distribution $b(v)$, the spectral energy distribution recorded is

$$b'(v) = a(v) * b(v) \qquad (10.1)$$

Two monochromatic lines are resolved when their separation $v_2 - v_1$ is greater than a certain value Δv which is determined by the width of $a(v)$. The resolving power of the spectroscope is then

$$\mathscr{R} = v/\Delta v = \lambda/\Delta\lambda \qquad (10.2)$$

The resolving power available with prisms and gratings is limited to about 10^6. Higher resolving powers can be achieved only by interference spectroscopy.

10.1. Étendue of an interferometer

An important characteristic of an interferometer, when it is used for the study of faint spectra, is its 'light-gathering power' or 'throughput', which is defined by an invariant known as its étendue. In the arrangement shown in

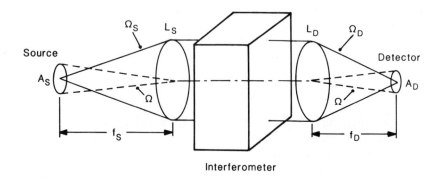

Fig. 10.1. Étendue of an interferometer.

Fig. 10.1, the effective areas A_S and A_D of the source and the detector are images of each other. These images subtend at the lenses L_S and L_D, respectively, the same solid angle

$$\Omega = A_S/f_S^2 = A_D/f_D^2 \qquad (10.3)$$

The amount of radiation accepted by the lens L_S is proportional to the area of the source and the solid angle Ω_S subtended by L_S at a point in the source. Similarly, the amount of radiation reaching the detector is proportional to the product of the area of the detector and the solid angle Ω_D subtended at a point in it by the lens L_D. Since the two lenses have the same effective area A,

$$\Omega_S = A/f_S^2 \qquad (10.4)$$

and

$$\Omega_D = A/f_D^2 \qquad (10.5)$$

Accordingly,

$$A_S\Omega_S = A_D\Omega_D = A\Omega = E \qquad (10.6)$$

where E is known as the étendue of the interferometer.

10.2. The Fabry–Perot interferometer

Among the instruments developed for direct observation of a spectrum at high resolution, Michelson's echelon and the Lummer–Gehrcke plate were widely used earlier [see Candler, 1951], but they have now been replaced completely by the Fabry–Perot interferometer (FPI). This instrument, which has been briefly discussed earlier in Section 4.1, usually consists of two fused silica plates which may have either a fixed or a variable separation. The inner faces of the plates are worked flat and are usually coated with highly reflecting multilayer dielectric stacks ($R > 0.95$). However, metal coatings are also used in applications such as wavelength comparisons where freedom from dispersion is necessary. The plates themselves are slightly wedged, so that the beams reflected from the outer surfaces can be easily eliminated by a suitably positioned aperture.

If the surfaces are separated by a distance d, the instrumental function of the FPI is the Airy function,

$$a(\nu) = T^2/[(1-R)^2 + 4R \sin^2 (\psi/2)] \tag{10.7}$$

where $\psi = (4\pi\nu nd/c) \cos \theta + \Delta\phi$, θ being the angle of incidence of a ray and $\Delta\phi$ the additional phase shift on reflection at the plates. This function, which is shown in Fig. 4.2, exhibits sharp peaks which are separated by a frequency difference $c/2nd$. This corresponds to the free spectral range which can be handled without successive orders overlapping. The half-width of the peaks $\Delta\nu$ is obtained by dividing the free spectral range by the finesse F, so that

$$\Delta\nu = (1/F)(c/2nd) \tag{10.8}$$

For an ideal FPI, the finesse is given by the relation (see Eq. 4.14)

$$F = \pi R^{1/2}/(1-R) \tag{10.9}$$

However, in any practical case, the finesse is always lower because of deviations of the surfaces from flatness and the finite aperture of the plates; it can be estimated from more precise formulas given by Ramsay [1969].

In one mode of operation, the spacing of the plates is fixed (the Fabry–Perot étalon). Each wavelength in the source then gives rise to a system of rings centred on the normal to the plates. Because of the limited free spectral range of the interferometer, it is necessary to separate the fringes for different spectral lines. This is done by imaging the rings on the slit of a spectrograph, as shown in Fig. 10.2, so that each line in the spectrum contains a narrow strip of the circular fringe pattern corresponding to that line. This technique has been used widely, since a single photograph of the spectrum can give information on the structure of a number of spectral lines.

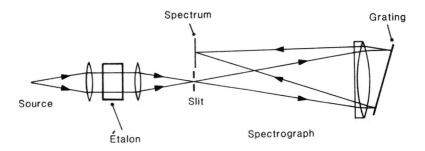

Fig. 10.2. Separation of the Fabry–Perot fringes for different spectral lines by a spectrograph.

10.3. The scanning Fabry–Perot spectrometer

Because of its circular symmetry, the FPI has a much larger étendue than the echelon and the Lummer–Gehrecke plate. However, in the configuration shown in Fig. 10.2, its étendue is limited by the narrow slit of the spectrograph. To obtain its full étendue, it is necessary to make use of all the rays having the same angle of incidence. For this reason, the FPI is now commonly used as a spectrometer. In this mode of operation, a small aperture is placed in the focal plane of a lens behind the FPI, and the transmitted intensity is recorded with a photomultiplier as the effective spacing of the plates is varied [Jacquinot and Dufour, 1948]. The size of this aperture is determined by $\mathscr{R}$ (the resolving power of the instrument), the maximum solid angle which the aperture can subtend at the source without significantly degrading the resolution being given by the relation

$$\Omega_m = 2\pi/\mathscr{R} \tag{10.10}$$

The FPI can be made to scan a selected region of the spectrum by changing either the spacing of the plates or the refractive index of the medium separating them. The latter system has the advantage of simplicity. Usually, the FPI is placed in an airtight box which is evacuated. Air is then allowed to enter at a controlled rate so that the pressure increases linearly with time. However, the spectral range which can be covered by pressure scanning is limited ($\Delta\lambda/\lambda = 3 \times 10^{-4}$ per atmosphere for air) and can only be increased by using higher pressures and denser gases [Jacquinot, 1960]. This limitation can be overcome with mechanical scanning. While a number of techniques have been described for this purpose, the most common method with spacings less than a few centimetres is to use piezoelectric spacers, parallelism of the plates being maintained by means of a servo system [Ramsay, 1966; Hicks *et al.*, 1974; Sandercock, 1976]. Systems have also been

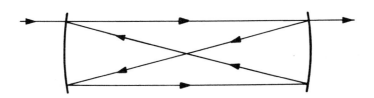

Fig. 10.3. Spherical Fabry–Perot interferometer.

described which utilize a microcomputer to ensure a strictly linear scan, as well as for the acquisition and storage of intensity data in a digital form [see, for example, Wood, 1978; Yamada *et al.*, 1980].

10.4. The spherical-mirror Fabry–Perot interferometer

It is apparent from Eqs (10.6) and (10.10) that, for plates of a given diameter, the étendue of a FPI with plane mirrors (plane FPI) is inversely proportional to its resolving power. This limitation is overcome in the spherical-mirror Fabry–Perot interferometer (spherical FPI) described by Connes [1956, 1958]. This consists, as shown in Fig. 10.3, of two spherical mirrors whose separation is equal to their radius of curvature r, so that their paraxial foci coincide, giving an afocal system. Any incident ray, after traversing the interferometer four times, falls back on itself, thus giving rise to an infinite number of outgoing rays which are coincident (and not just parallel, as with the plane FPI), with an increment in optical path $\Delta p = 4r$ between successive rays.

The étendue of the spherical FPI is limited only by the spherical aberration of the mirrors, which restricts their useful aperture. This restriction is less serious with high resolution intruments in which the radius of curvature of the mirrors is much greater than their diameter, and the superiority of the spherical FPI over the plane FPI is most marked under these conditions. This is apparent from a comparison of a plane FPI of diameter D and thickness d with a spherical FPI having mirrors of the same aberration-limited diameter and a separation $r = d/2$. Both these interferometers have the same free spectral range and, if the reflectances of the mirrors are properly chosen, the same finesse and, hence the same resolving power. However, the spherical FPI exhibits a gain in étendue over the plane FPI

$$M = 2(d/D)^2 \tag{10.11}$$

This gain is quite high when the separation of the mirrors is large compared to their diameter.

A further advantage of the spherical FPI is that, once the separation of the mirrors has been set, any angular misalignment merely redefines the optical axis of the system. With the plane FPI, a similar misalignment is equivalent to an imperfection in the surfaces of the mirrors.

A spherical FPI is usually operated as a scanning instrument [Herriot, 1963; Fork et al., 1964]. While it has the disadvantage of a fixed free-spectral range, this is more than offset by its increased étendue and ease of adjustment. The optimization of its design has been discussed in detail by Hercher [1968]. Scanning spherical FPI's are now used quite commonly to examine the pattern of longitudinal and transverse modes in the output of a cw laser. With pulsed lasers, where the FPI cannot be operated in a scanning mode, the mirror separation can be made slightly less than the common radius of curvature. Such a defocused spherical FPI produces a pattern of concentric rings with quasi-linear spectral dispersion which can be photographed conveniently [Bradley and Mitchell, 1968].

10.5. The multiple Fabry–Perot interferometer

To study a complex spectrum with a scanning FPI, it is necessary to use a monochromator which can supply the FPI with a narrow enough spectral band to avoid overlapping of successive orders. If the étendue of the combination is not to be limited by the monochromator, it must have an étendue equal to or greater than that of the FPI. This condition cannot be met with a conventional grating or prism monochromator, but can be satisfied by using two or more FPI's in series [Chabbal, 1958]. If multiple reflections between the FPI's are eliminated, so that the FPI's are effectively decoupled (either by tilting them slightly so that they are not parallel, or by introducing a neutral density filter between them), the overall instrumental function is simply the product of the instrumental functions for the individual FPI's.

In such an arrangement, the second FPI serves to eliminate the unwanted transmission peaks of the first FPI over a relatively wide range. The free spectral range can be increased by a factor m (typically $m \leqslant 10$) simply by making the separation of the plates in the second FPI $(1/m)$ times that in the first. However, an alternative which gives a narrower half-width and higher efficiency is to make the spacing of the second FPI $(m+1)/m$ times that of the first [Mack et al., 1963].

To explore a spectrum, the pass bands of all the FPI's must be shifted simultaneously. A simple way of doing this is by pressure scanning, maintaining a constant pressure difference between the individual FPI's [Mack et al., 1963; Ramsay et al., 1970]. More complex systems with a computer-controlled scan have also been described [Winter, 1984].

10.6. The multiple-pass Fabry-Perot interferometer

The contrast factor of a FPI, defined by the ratio of the intensities of the maxima and the minima in the instrumental function, is

$$C = [(1+R)/(1-R)]^2 \tag{10.12}$$

where R is the reflectance of the mirrors. With typical coatings ($R \approx 0.95$) this figure is not adequate to ensure that a weak satellite is not masked by the background due to a neighbouring strong spectral line. A much higher contrast factor, close to the square of that given by Eq. (10.12), can be obtained by passing the light twice through the same interferometer [Dufour, 1951; Hariharan and Sen, 1961f]. This technique has been applied very successfully to Brillouin spectroscopy by Sandercock [1970], and resolving powers of 10^8, with contrast factors greater than 10^{10} have been obtained in scanning FPI's using prism combinations to give up to five passes [Lindsay and Shepherd, 1977; Vacher et al., 1980].

10.7. Birefringent filters

An interferometer consisting of a plate of a birefringent material such as quartz or calcite, with a thickness d, cut with its faces parallel to the optic axis and set between parallel polarizers with its optic axis at 45°, has a transmission function

$$a(v) = (1/2) \cos^2 \pi v T \tag{10.13}$$

where $T = (n_o - n_e)d/c$ is the delay introduced in the interferometer. A single transmission peak can be obtained by using a number of such interferometers in series, each one with a delay twice that of the preceeding one. This is the principle of the birefringent filter developed by Lyot [1944] and, independently, by Öhman [1938] for studies of the solar surface. The filter can be made tunable over one order and a wider field obtained by splitting each element into two plates of equal thickness with their axes crossed, and interposing between them a rotatable half-wave plate between two fixed quarter-wave plates. A detailed description of such a filter with a pass band only 0.0125 nm wide has been given by Steel et al. [1961].

An alternative type of birefringent filter described by Šolc [1965] consists of m plates with equal retardations set between two polarizers. In one (fan) form, the polarizers are set parallel and the plates have their optic axes at angles $\pi/4m, 3\pi/4m, \ldots (2m-1)\pi/4m$ to that of the polarizers, while in

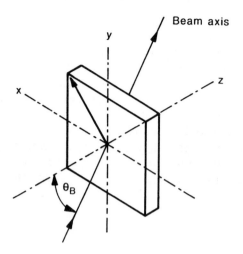

Fig. 10.4. Birefringent tuning element for a dye laser.

another (folded) form the polarizers are crossed, and alternate plates are set at angles of $+\pi/4m$ and $-\pi/4m$.

Detailed theories of both these types of birefringent filters have been given by Harris *et al.* [1961] and Ammann and Chang [1965]. These permit the design of filters with any desired periodic transmittance.

A major application of such birefringent filters is as wavelength-selection elements in the resonators of tunable dye lasers [Yarborough and Hobart, 1973; Bloom, 1974]. Prisms and gratings have the disadvantage that a large beam diameter is required to attain high resolution, while étalons have a very limited tuning range. These problems are avoided with a birefringent filter.

The simplest arrangement for such a wavelength selector consists, as shown in Fig. 10.4, of a birefringent plate cut with its faces parallel to the optic axis and mounted in the laser cavity so that the beam is incident on it at the Brewster angle θ_B. If the optic axis does not lie in the plane of polarization of the laser beam (the yz plane) the transmittance of the plate is unity only for wavelengths satisfying the condition

$$[(n_0 - n_e)d \sin^2 \eta]/\sin \theta_B = m\lambda \tag{10.14}$$

where n_o and n_e are the ordinary and extraordinary refractive indices of the plate, d is its thickness, η is the angle between the ray in the crystal and the optic axis and m is an integer. At other wavelengths the polarization of the beam is modified, as a result of which it suffers increased losses due to reflection at the surface of the plate. The filter can be tuned by rotating the plate in its own plane. This changes the angle between the optic axis and the axis of the laser beam and, hence, the angle η in Eq. (10.14).

The bandwidth of such a filter consisting of a stack of plates (usually three) whose thickness are in integer ratios is narrower than that of a Lyot filter with the same free spectral range. Even though the transmittance does not go to zero at the minima, the drop in transmittance is enough to ensure that the laser oscillates at only one frequency. Increased discrimination against secondary peaks can be obtained by adding glass plates at the Brewster angle [Holtom and Teschke, 1974].

A systematic design procedure which yields a filter with the maximum tuning range and the minimum number of plates has been described by Hodgkinson and Vukusic [1978]. Typically, a bandwidth of 0.025 nm and a tuning range of 100 nm can be obtained with a three-element filter.

10.8. Laser wavelength meters

The increasing use of tunable dye lasers in spectroscopy has led to the need for an instrument which can be used to measure their output wavelength, in real time, with an accuracy commensurate with their linewidth (< 1 part in 10^8).

Interference wavemeters can be divided into two broad categories, dynamic wavemeters, in which the measurement involves the movement of some element or group of elements, and static wavemeters, which have no moving parts (other than those needed for initial alignment of the instrument). Dynamic wavemeters offer the advantage of greater accuracy, but can only be used with cw lasers, while static wavemeters can also be used with pulsed lasers.

10.9. Dynamic wavelength meters

The most common type of dynamic wavelength meter uses a two-beam interferometer in which the number of fringes crossing the field is counted as the optical path is changed by a known amount. One form [Kowalski et al., 1976] is shown in Fig. 10.5. In this, two beams, one from the dye laser whose wavelength λ_1 is to be determined and the other from a reference laser (a frequency-stabilized He–Ne laser) whose wavelength λ_2 is known, traverse the same two paths in opposite directions. Another form [Hall and Lee, 1976] uses a folded Michelson interferometer in which the two end reflectors are mounted back to back on a single carriage, so that, when the carriage is moved, one optical path increases while the other decreases by the same amount. In both instruments, the beams from the dye laser and the reference

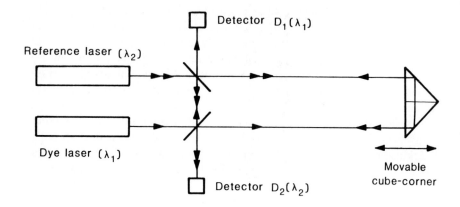

Fig. 10.5. Dynamic wavelength meter [Kowalski *et al.*, 1976].

laser emerge separately, and the fringe systems formed by the two wavelengths are imaged on separate detectors.

If, in the interferometer shown in Fig. 10.5, the end reflector is moved through a distance d, the number of fringes seen by the detector D_1 is

$$N_1 = 2n_1 d/\lambda_1 \qquad (10.15)$$

where n_1 is the refractive index of air at the wavelength λ_1. Similarly, the number of fringes seen by the detector D_2 is

$$N_2 = 2n_2 d/\lambda_2 \qquad (10.16)$$

where n_2 is the refractive index of air at the wavelength λ_2. Accordingly,

$$\lambda_1 = (N_2 n_1/N_1 n_2)\lambda_2 \qquad (10.17)$$

If the interferometer is operated in a vacuum, the ratio (n_1/n_2) is unity, and the wavelength of the dye laser can be determined directly from Eq. (10.17) by counting fringes simultaneously at both wavelengths. If the interferometer is operated at normal atmospheric pressure, a correction must be made for the ratio (n_1/n_2) which can be calculated from formulas established by Edlén [1966].

A convenient feature of these instruments is that the result is independent of the velocity of the end reflectors. In its simplest form, in which only integer changes in the order number are counted, the precision is approximately $1/N_2$, where N_2 is the total change in the integer order for the reference laser.

Typically, for a change in the optical path difference of one metre, $N_2 \approx 1.6 \times 10^6$ giving a precision of about 7 parts in 10^7.

Higher precision can be obtained by phase-locking an oscillator to an exact multiple m of the frequency of the ac signal from the reference channel and measuring the frequency of this oscillator. This allows the fractional order number for the reference laser to be determined to $\pm 1/m$ [Hall and Lee, 1976; Kowalski et al., 1978]. Higher resolution can also be obtained by digital averaging of the two modulation frequencies. This allows interpolation of the fractional order number to $N_2^{-\frac{1}{2}}$. With careful design to ensure a constant velocity of the reflectors, a precision of the order of 1 part in 10^9, with a measuring time of a few seconds, is possible, though the acuracy may be less because of systematic errors [Monchalin et al., 1981].

A more compact design for a dynamic wavemeter is based on a scanning spherical FPI [Salimbeni and Pole, 1980]. If the separation of the mirrors in such an interferometer is changed while it is illuminated by a single wavelength, the transmitted intensity consists of a series of pulses, each pulse corresponding to a change in the interference order of unity. With two lasers, coincidences will occur between the transmission peaks corresponding to the two wavelengths at intervals which satisfy the condition

$$m_1\lambda_1 = m_2\lambda_2 = p \qquad (10.18)$$

where m_1 and m_2 are the changes in the integer orders and p is the change in the optical path difference. If the output pulses due to the two wavelengths are detected separately and counted over this interval, the wavelength of the dye laser can be calculated from Eq. (10.18). In this arrangement the precision is enhanced by a factor equal to the finesse of the FPI over that obtained by fringe counting over the same range in a Michelson interferometer. However, defocusing (see Section 10.4) limits the range of movement of the mirrors. A precision of 1 part in 10^7 is possible with a range of movement of 25 mm, in a time of 0.8 s.

10.10. Static wavelength meters

A number of static wavelength meters have been developed based on the Michelson, Fabry–Perot and Fizeau interferometers.

The sigma-meter [Juncar and Pinard, 1975, 1982] uses four Michelson interferometers with optical path differences of 0.5 mm, 5 mm, 50 mm and 500 mm sharing a common beam-splitter and reference mirror. A prism in the reference arm acts as an achromatic $\lambda/4$ plate to produce two orthogonally polarized reference waves with a phase difference of 90°. These give

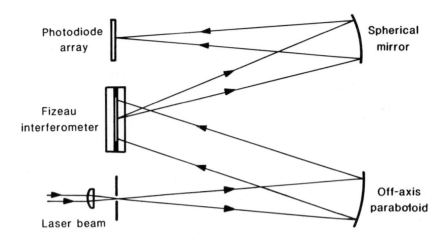

Fig. 10.6. Wavelength meter using a Fizeau interferometer.

rise to two separate fringe systems in each interferometer, in which the intensity distributions are, respectively,

$$I_1 = I_0 (1 + \cos 2\pi\nu\tau) \qquad (10.19)$$

and

$$I_2 = I_0 (1 + \sin 2\pi\nu\tau) \qquad (10.20)$$

so that the fractional fringe order can be determined from these intensities. If the wavelength of the dye laser is known approximately from measurements with a monochromator, the integer order of the first interferometer can be calculated. The wavelength determined with this interferometer is then used to calculate the integer order for the next interferometer. This process is repeated successively to obtain the interference order with the last interferometer and, hence, the actual wavelength. An accuracy of 1 part in 10^8 is possible with this instrument.

Another type of static wavemeter uses three FPI's with free spectral ranges of 3 GHz, 30 GHz and 300 GHz, respectively, along with a monochromator as a pre-filter [Byer *et al.*, 1977; Fischer *et al.*, 1981]. In this instrument, the intensity distributions in the fringes are read simultaneously by three photodiode arrays, and the fractional order for each interferometer is calculated with an accuracy of 0.01 by a minicomputer. As with the sigma-meter, the approximate wavelength is given by the setting of the monochromator, and the final wavelength is found by a sequence of approximations, using each of the interferometers in turn. This instrument has a potential precision of about 5 parts in 10^9.

One of the simplest static wavemeters is that developed by Snyder [1977] which uses a single Fizeau interferometer. This consists, as shown in Fig. 10.6, of two uncoated fused-silica optical flats, about 1 mm apart, making a small angle ϵ ($\epsilon \approx 3$ minutes of arc) with each other. Light from the dye laser is collimated by an off-axis parabolic mirror and used to illuminate the interferometer. Fringes of equal thickness are then formed, which run parallel to the apex of the wedge. The intensity distribution in these fringes, along a line perpendicular to the apex of the wedge, is given by the relation

$$I(x) = I_0[1 + \cos(2\pi x/\lambda\epsilon)] \qquad (10.21)$$

This intensity distribution is recorded directly by a 1024-element, linear photodiode array which receives the light reflected from the interferometer. The spatial period of the fringe pattern can then be used to evaluate the integer part of the interference order, while the fractional part can be calculated from the positions of the minima and the maxima with respect to a reference point on the wedge.

Errors due to imperfect collimation as well as small changes in the angle of incidence can be minimized in this arrangement by positioning the detector array so that the shear between the beams reflected from the two surfaces of the wedge is zero at the surface of the detector [Snyder, 1981; Gardner, 1983]. A precision of the order of 1 part in 10^7 with an update rate of 15 Hz is possible with this instrument.

10.11. Heterodyne techniques

Heterodyne techniques can also be used for very precise measurements of the frequency of a laser and, hence, its wavelength. In one experiment [Bay et al., 1972] electro-optic modulation of a laser beam (frequency ν_L) at a microwave frequency ν_M ($\approx 10^{10}$ Hz) produced two sideband frequencies $\nu_L \pm \nu_M$. The length L of a Fabry–Perot interferometer as well as the modulation frequency ν_M were then adjusted simultaneously so that both sidebands were transmitted by the interferometer with maximum intensity. Under these conditions it can be shown that

$$\nu_L = (N/\Delta N)2\nu_M \qquad (10.22)$$

where N is the order of interference for the frequency ν_L, and ΔN is the difference in the interference orders for the two sidebands. This technique directly gave the frequency of the laser to a few parts in 10^8.

Another technique for comparing the frequencies of two lasers [Baird, 1983b] makes use of two slave lasers which can be tuned around the frequencies of the two lasers. These slave lasers are locked to two transmission peaks of a very stable Fabry–Perot interferometer so that their wavelengths are integral submultiples of the optical path difference in the interferometer. The frequency offsets of the lasers whose wavelengths are to be compared are then determined by measuring the beat frequency produced by each laser with the corresponding slave laser.

11
Fourier-transform
spectroscopy

As outlined in Sections 3.6 and 3.11, the spectral energy distribution of a source can be obtained from measurements of the complex degree of coherence in a two-beam interferometer, as the path difference between the interfering wavefronts is varied.

Michelson made use of this fact when he inferred the structure of several spectral lines from a study of the visibility curves. However, measurements of the visibility only give the modulus of the Fourier transform of the spectral energy distribution. A unique spectral distribution can be obtained from the visibility curve only when the spectral distribution is symmetrical with respect to a central frequency.

In Fourier-transform spectroscopy (FTS), the intensity at a point in the interference pattern is recorded as a function of the delay in the interferometer (the interference function). The variable part of the interference function (called the interferogram), on Fourier transformation, then gives the spectrum without any such ambiguity. One of the earliest spectra obtained in this manner is shown in Fig. 11.1, along with the corresponding interferogram [Fellgett, 1958]. This technique has been described in detail in a number of reviews [see Vanasse and Sakai, 1967; Chamberlain, 1972] and is now used very widely; two reasons for this are the étendue and multiplex advantages.

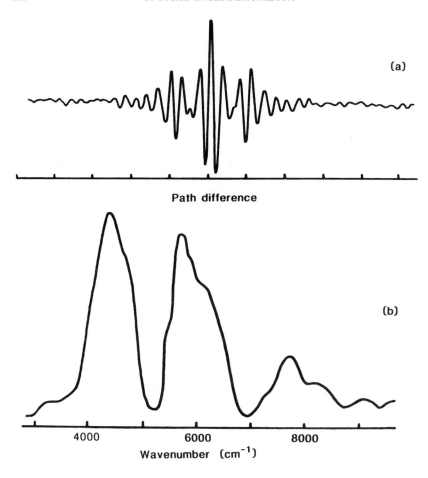

Fig. 11.1.   (a) Interferogram and (b) spectrum obtained with the 3.5 magnitude star, α Herculis [Fellgett, 1958].

11.1. The étendue and multiplex advantages

An interferometer in which the interference pattern exhibits radial symmetry, such as the FPI, has the advantage over a prism or grating spectrograph of a much greater étendue (the étendue or Jacquinot advantage, see Section 10.3). However, when such an interferometer is operated in the scanning mode to make full use of the available étendue, the total scanning time T is divided between, say, m elements of the spectrum, so that each element is observed only for a time equal to (T/m). If the main source of noise is the detector, which is often the case in the far infrared, the noise power is independent of the signal. This results in a reduction of the signal-to-noise (S/N) ratio by a factor $m^{1/2}$, compared to the situation if each spectral element were recorded over the full time T.

The Michelson interferometer has the same étendue advantage as the FPI. In addition, when the delay is made to vary linearly with time, each element of the spectrum gives rise to an output which is modulated at a frequency inversely proportional to its wavelength. It is therefore possible to record all these signals simultaneously (or, in other words, to multiplex them) and decode them later to obtain the spectrum. Since each spectral element is now recorded over the full scan time T, an improvement in the S/N ratio by a factor of $m^{1/2}$ over a conventional scanning instrument is obtained. This is known as the multiplex (or Fellgett) advantage.

11.2. Theory

The main component of a Fourier transform spectrometer is usually, as shown in Fig. 11.2, a Michelson interferometer illuminated with an approximately collimated beam. With monochromatic light, the output of the detector is

$$G(\tau) = g(\nu) (1 + \cos 2\pi\nu\tau)$$
$$= g(\nu) + g(\nu) \cos 2\pi\nu\tau \tag{11.1}$$

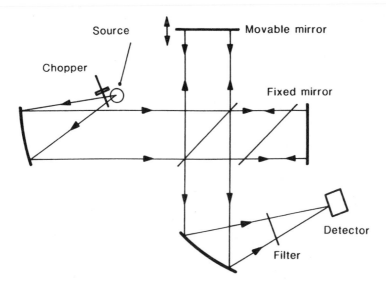

Fig. 11.2. Michelson interferometer adapted for Fourier-transform spectroscopy.

where τ is the delay, and

$$g(\nu) = I(\nu)T(\nu)d(\nu) \tag{11.2}$$

is the product of three spectral distributions: the radiation studied $I(\nu)$, the transmittance of the spectroscope $T(\nu)$, and the detector sensitivity $d(\nu)$. When τ varies linearly with time, $G(\tau)$ exhibits a cosinusoidal variation of constant amplitude extending to $\tau = \pm \infty$.

If, however, the source has a large spectral bandwidth, the output of the detector is

$$G(\tau) = \int_0^\infty g(\nu)d\nu + \int_0^\infty g(\nu) \cos 2\pi\nu\tau \, d\nu \tag{11.3}$$

Since the first term on the right hand side of Eq. (11.3) is a constant, the variable part of the output, which constitutes the interferogram, is

$$F(\tau) = \int_0^\infty g(\nu) \cos 2\pi\nu\tau \, d\nu \tag{11.4}$$

The interferogram initially exhibits large fluctuations when $|\tau|$ is increased from 0, since all the spectral components are in phase when $\tau = 0$, but the amplitude of these fluctuations drops off rapidly as $|\tau|$ increases.

The actual spectral distribution $g(\nu)$ exists only for $\nu > 0$. However, for convenience, we can represent this asymmetrical function, as shown in Fig. 11.3, as the sum of an even component $g_e(\nu)$ defined by the relation

$$g_e(\nu) = g_e(-\nu) \tag{11.5}$$

and an odd component $g_o(\nu)$ defined by the relation

$$g_o(\nu) = -g_o(-\nu) \tag{11.6}$$

so that

$$g(\nu) = g_e(\nu) + g_o(\nu) \tag{11.7}$$

In addition, since the two components are numerically the same,

$$|g_e(\nu)| = |g_o(\nu)| \tag{11.8}$$

Accordingly, Eq. (11.4) can be rewritten as

$$F(\tau) = \int_{-\infty}^\infty [g_e(\nu) + g_o(\nu)]\cos 2\pi\nu\tau \, d\nu$$

$$= 2 \int_{-\infty}^\infty g_e(\nu) \cos 2\pi\nu\tau \, d\nu \tag{11.9}$$

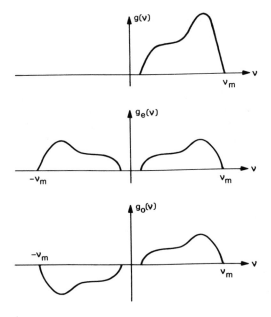

Fig. 11.3. Decomposition of the physical spectrum $g(\nu)$ into an even component $g_e(\nu)$ and an odd component $g_o(\nu)$.

Fourier inversion of Eq. (11.9) then gives

$$g_e(\nu) = 2 \int_0^\infty F(\tau) \cos 2\pi\nu\tau \, d\tau \qquad (11.10)$$

However, if we consider only positive values of ν, $g_e(\nu)$ and $g_o(\nu)$ are identical and their sum is equal to $g(\nu)$. Hence, over this range ($\nu > 0$),

$$g_e(\nu) = (1/2)g(\nu) \qquad (11.11)$$

so that

$$g(\nu) = 4 \int_0^\infty F(\tau) \cos 2\pi\nu\tau \, d\tau \qquad (11.12)$$

11.3. Resolution and apodization

The calculation of $g(\nu)$ from Eqs. (11.10) and (11.12) requires a knowledge of the interferogram over an infinite range of values of τ, but, in any practical case, the range of the interferogram is limited by the distance over which the end mirror of the interferometer can be moved. The effect of finite mirror travel can be taken into account by rewriting Eq. (11.10) as

$$g'_e(\nu) = 2 \int_0^\infty A(\tau)F(\tau) \cos 2\pi\nu\tau \, d\tau \qquad (11.13)$$

where $A(\tau)$ is a function which drops to zero at the maximum allowed value of τ. In the frequency domain, Eq. (11.13) becomes

$$g'_e(\nu) = a(\nu) * g_e(\nu) \qquad (11.14)$$

where

$$a(\nu) \leftrightarrow A(\tau) \qquad (11.15)$$

A comparison of Eq. (11.15) with Eq. (10.1) shows that $a(\nu)$ is the instrumental function of the spectrometer.

For the simplest case in which the interferogram is truncated at $\pm\tau_m$, $A(\tau)$ is the rectangular function

$$A(\tau) = \begin{cases} 1, & \text{when } |\tau| \leq \tau_m \\ 0, & \text{when } |\tau| > \tau_m \end{cases} \qquad (11.16)$$

and the instrumental function is

$$a(\nu) = \text{sinc} \, (2\nu\tau_m) \qquad (11.17)$$

If we take the resolution limit $\Delta\nu$ to be the separation of the first zero of this function from its peak, we have

$$\Delta\nu = 1/2\tau_m \qquad (11.18)$$

The sinc function is not a desirable instrumental function, since it does not drop smoothly to zero and the sidelobes due to a strong spectral line could be mistaken for other weak spectral lines. These sidelobes can be eliminated by a process known as apodization [Jacquinot and Roizen-Dossier, 1964], which involves multiplying the interferogram with a weighting function which progressively reduces the contribution of greater delays and thereby eliminates the sharp cut-off at $\tau = \tau_m$. A linear taper is commonly used, in which case

$$A(\tau) = \begin{cases} 1 - |\tau|/\tau_m, & |\tau| \leq \tau_m \\ 0, & |\tau| > \tau_m \end{cases} \qquad (11.19)$$

and

$$a(\nu) = \mathrm{sinc}^2(\nu \tau_m) \tag{11.20}$$

The side lobes are reduced appreciably, but at the expense of a doubling of the width of the instrumental function, and a corresponding loss in resolving power. Other apodization functions and their effects on the instrumental function have been reviewed by Vanasse and Sakai [1967].

11.4. Sampling

When calculating the Fourier transform by means of a digital computer it is necessary to know the minimum number of points at which the interferogram must be sampled as well as the minimum number of points for which the transform must be computed to recover an undistorted spectrum. This problem was analysed by Strong and Vanasse [1959] and, later, by Connes [1961]. We make use of the fact that if an interferogram taken over a range of delays from 0 to τ_m is sampled at M points, that is to say with a sampling interval τ_m / M, the recovered spectrum is repeated, as shown in Fig. 11.4, at a frequency interval M/τ_m. For a physical spectral distribution covering the range of frequencies 0 to ν_m, the recovered spectrum $g_e(\nu)$, defined by Eq. (11.10), has a spectral bandwidth of $2\nu_m$. Accordingly, to ensure that there is no overlap (aliasing) of successive repetitions of the recovered spectrum, we must have

$$2\nu_m \leqslant M/\tau_m \tag{11.21}$$

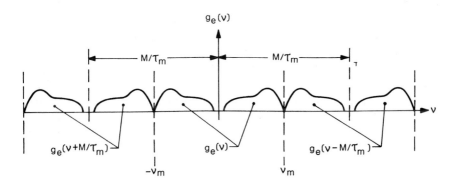

Fig. 11.4. Repetition (aliasing) of the recovered spectrum $g_e(\nu)$ when the interferogram is sampled at an interval $\tau_m/M = 1/2\nu_m$.

so that the minimum number of points at which the interferogram is sampled is

$$M = 2\nu_m \tau_m \tag{11.22}$$

In the spectral domain, the number of points N for which the transform must be computed is given by the bandwidth ν_m divided by the resolution limit $\Delta\nu$. From Eq. (11.18),

$$\begin{aligned} N &= \nu_m / \Delta\nu \\ &= 2\nu_m \tau_m \end{aligned} \tag{11.23}$$

Hence, from Eq. (11.22), N is also equal to M.

11.5. Effect of the finite areas of the source and the detector

Equation (11.18), which defines the resolution limit, is valid only when the interferometer is illuminated with a perfectly collimated beam. If the source and the detector subtend a finite angle 2α at the entrance pupil and the exit pupil respectively (see Fig. 10.1), there will be an additional phase difference between the axial ray and a limiting off-axis ray

$$\begin{aligned} \Delta\phi &= 2\pi\nu\tau(1 - \cos\alpha) \\ &\approx \pi\nu\tau\alpha^2 \\ &\approx \nu\tau\Omega \end{aligned} \tag{11.24}$$

since $\Omega = \pi\alpha^2$. As a result, the calculated frequency ν' is no longer the true frequency ν, but is given by the relation

$$\nu' = \nu(1 - \Omega/4\pi) \tag{11.25}$$

In addition, there is a progressive decrease in the visibility $\mathscr{V}$ of the fringes, with increasing delay. It can be shown [Terrien, 1958] that

$$\mathscr{V}(\tau) = \text{sinc } (\Omega\nu\tau/2\pi) \tag{11.26}$$

so that the fringes disappear when

$$\Omega = 2\pi/\nu\tau \tag{11.27}$$

The maximum solid angle which the source and the detector can subtend at the collimating optics is then, from Eqs (11.27) and (11.18),

$$\Omega_m = 2\pi/\mathscr{R} \tag{11.28}$$

where $\mathscr{R} = \nu/\Delta\nu$ is the resolving power.

11.6. Field widening

It is possible to increase Ω_m beyond the limit given by Eq. (11.28) by using a system in which, as in the spherical FPI, emerging rays derived from the same incident ray are coincident and not merely parallel. This permits the delay to be increased while keeping the shift (see Section 3.9) zero. Systems for this purpose have been described by Connes [1956] and by Bouchareine and Connes [1963]. Such systems can give an increase of étendue by as much as 100 but have other drawbacks which limit their use.

11.7. Phase correction

If the interferometer is not perfectly compensated (or, in other words, has residual dispersion in one path), Eq. (11.4) is no longer valid and the interferogram is represented by the relation

$$F_a(\tau) = \int_0^{\nu_m} g(\nu) \cos\left[2\pi\nu\tau + \phi(\nu)\right]d\nu \tag{11.29}$$

In such a case, the modulus $g_e(\nu)$ can be recovered by taking the absolute value of the Fourier transform of $F_a(\tau)$. A better method is to generate the required symmetric interferogram from the measured unsymmetric interferogram by convolution with a phase correction function. Since the phase error is a smooth, slowly varying function it can be computed separately from a short, two-sided interferogram [Forman et al., 1966].

11.8. Noise

Noise can be classified in two categories: noise arising in the detector or interferometer, which is added to the signal (additive noise); and noise associated with the incident radiation, whose power varies with the signal level (multiplicative noise). The latter can again be subdivided into photon noise and signal noise. Photon noise arises from the random fluctuations in the number of photons arriving in a given time interval and is therefore proportional to the square root of the signal; signal noise, on the other hand, arises from fluctuations in the transmittance of the medium between the source and the detector, or from fluctuations in the output of the source itself, and is, therefore, proportional to the signal.

As shown in Section 11.1, the multiplex advantage applies for additive noise, which is independent of the signal power. With photon noise, it can be shown that multiplexing results in no advantage, while with signal noise, multiplexing is at a disadvantage when compared to sequential measurement. Accordingly, the multiplex advantage disappears in the visible and ultraviolet regions of the spectrum, where the noise is largely either photon noise or signal noise. Fourier-transform spectroscopy is therefore most useful in the far infrared, where the energy of individual photons is low and detector noise is frequently the limitation.

11.9. Pre-filtering

With a source emitting over a wide spectral band, the dynamic range required to record an interferogram is very large. This is because all the frequencies contribute to a grand maximum at zero path difference, while relatively small variations of power at large path differences correspond to weak absorption lines. To ensure that information on such weak features is not lost, the dynamic range of the recording system must be greater than the S/N ratio in the interferogram. If this condition cannot be met, it is necessary to use optical filters to limit the spectral bandwidth studied and thereby reduce the dynamic range required.

Dynamic range is not so much of a problem now, since digital recording systems with a discrimination better than 1 part in 10^5 are available. However, even with such systems, a problem which can be encountered is aliased noise.

If the detector is sensitive up to a frequency ν_D which is higher than the maximum frequency ν_{max} up to which the spectrum is to be computed, the interferogram will contain a significant amount of photon noise and signal noise contributed by higher frequencies which are not of interest. When the interferogram is sampled and Fourier transformed, noise belonging to the adjacent repetitions of the spectrum, or aliased noise (see Fig. 11.4), will then overlap the recovered spectrum and degrade it.

Aliased noise can be eliminated, in principle, by the use of a filter which has unit transmittance between frequencies of ν_{min} and ν_{max} and zero transmission elsewhere. The Fourier transform of such a filter is a sinc function. Accordingly, if the original interferogram is convolved with this sinc function before it is transformed, all noise at frequencies outside the range of interest will be eliminated and cannot be aliased back. This procedure is called mathematical filtering [Connes and Nozal, 1961].

11.10. Interferometers for Fourier-transform spectroscopy

The most widely used interferometer for FTS is the Michelson inter-ferometer. In the near infrared, thin films of Ge or Si on BaF or CaF_2 plates are used as beam splitters. Because of the difficulty of finding suitable beam splitters which could be used in the far infrared, Strong and Vanasse [1958, 1960] developed an alternative in the lamellar-grating interferometer. This consists essentially of two mirrors broken up into strips so that one mirror can pass through the other. However, this interferometer must be used with a slit source narrow enough to ensure that adjacent strips are coherently illuminated; in addition, at longer wavelengths, the delay is determined by the velocity of the radiation in the slots in the mirrors. Subsequently, it was shown that a thin film of Mylar could be used as a beam-splitter in the far infrared; wire-mesh beam-splitters have also been used.

In either case, the slide carrying the moving mirror must be of very high quality to avoid tilting of the mirror; this problem can be minimized in the Michelson interferometer by replacing the mirrors by cube-corners or by 'cat's-eye' reflectors [Connes and Connes, 1966].

Two approaches to the movement of the mirror (scanning) have been followed. In one, known as periodic generation, the mirror is moved repeat-edly over the desired scanning range at a sufficiently rapid rate that the fluctuations of the output due to the passage of the fringes occur at an audio frequency. This has the advantage that ac amplification of the signal is possible without any additional modulation. A number of values of the output obtained for each value of the delay are then averaged digitally to give the final interferogram. In the other approach, known as aperiodic generation, the mirror is moved only once, relatively slowly, over the scan-ning range and the detector output is recorded at regular intervals. In this mode, the signal level is essentially steady during each observation.

Aperiodic generation is possible either with a continuous movement of the mirror or with a stepped movement. In the former case, the average value of the detector output is recorded during a brief time interval centred on each sampling point. This has the advantage that the drive system can be very simple. In the latter case, the mirror remains stationary during the sampling time and then moves rapidly to the next position, thus making maximum use of the observing time.

With either system of aperiodic generation, it is necessary to use some form of flux modulation so that ac amplification and synchronous detection can be used. The earliest spectrometers used amplitude modulation, usually by means of a chopper, but a better method is phase modulation [Chamberlain, 1971; Chamberlain and Gebbie, 1971]. In this technique, the mirror which usually is fixed is made to oscillate to and fro at a frequency

f so that, at any position of the stepped mirror, the delay varies with time according to the relation

$$\tau(t) = \tau + \Delta\tau \cos 2\pi f t \qquad (11.30)$$

For any input frequency ν, the output from the detector is then

$$I(t) = I_0 \{1 + \cos 2\pi\nu[\tau + \Delta\tau \cos 2\pi f t]\} \qquad (11.31)$$

The right-hand side of Eq. (11.31) can be expressed as the sum of a number of harmonics; with a filter that passes only the fundamental, the output is

$$I'(t) = 2I_0 J_1(2\pi\nu\Delta\tau) \sin 2\pi\nu\tau \qquad (11.32)$$

Phase modulation has the advantage over amplitude modulation that, since the beam is not interrupted by the chopper, the reduction in output is minimal.

11.11. Computation of the spectrum

The total number of operations involved in computing a Fourier transform by conventional routines is approximately $2M^2$ where M is the number of points at which the interferogram is sampled. When M is large, this leads to quite long computing times. Because of this, there were several early experiments in the use of analogue techniques. However, digital computing has now become so much faster and cheaper that these early attempts are only of historic interest. In addition, the fast Fourier transform (FFT) algorithm [Cooley and Tukey, 1965; Forman, 1966] has reduced the number of operations to $3M \log_2 M$. In an extreme case, when $M \approx 10^6$, the reduction in computing time is about 25 000, so that the computation of even very complex spectra becomes feasible. Procedures for computation have been discussed in detail by Bell [1972].

11.12. Applications

Rapid scanning FTS has found many applications in infrared spectroscopy; these include emission spectra and chemiluminescence, absorption spectra of aqueous solutions and transient species and studies of the kinetics of chemical reactions [Bates, 1976].

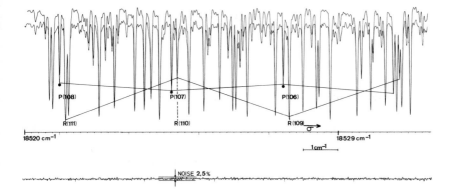

Fig. 11.5. A small section of the absorption spectrum of iodine in the visible ($\lambda \approx 540$ nm) obtained with a Fourier transform spectrometer (effective resolving power about 5×10^5). The two independent traces demonstrate the reproducibility and low noise of the method [Luc and Gerstenkorn, 1978].

In addition, because of their high étendue, Fourier transform spectrometers can be used to record spectra from very faint sources. For this reason they have been extensively used in studies of planetary atmospheres [Connes and Connes, 1966; see also Hanel and Kunde, 1975] and night sky emission [Baker *et al.*, 1973]. Because of their high resolving power, they have also been used in studies of the molecular spectra of gases at low pressures [Guelachvili, 1978]. Even though the multiplex advantage does not hold in the visible region, Fourier transform spectroscopy retains the advantages over a prism or grating of high resolving power and high accuracy of measurement of wavelengths, and has been found very useful for the study of complex molecular spectra [Luc and Gerstenkorn, 1978]. Figure 11.5, which shows a small section of the absorption spectrum of iodine obtained with a Fourier transform spectrometer having an effective resolving power of 5×10^5, gives an idea of the reproducibility and low noise possible with this method.

12
Holography

Holography permits complete reconstruction of a wavefront, that is to say, reproduction of the relative phases as well as the relative amplitudes of the light waves scattered by an object. This cannot be done with conventional imaging techniques, since all available recording materials respond only to the intensity, and information on the phase is lost. Gabor [1948, 1949, 1951] solved this problem by an artifice which has its roots in interferometry; namely, the addition of a coherent background. As a result, the phase variations across the wavefront are encoded as variations in the position of the resulting fringes.

The techniques of holography have been described in detail [see, for example, Collier *et al.*, 1971; Caulfield, 1979; and Hariharan, 1984]. Accordingly, this chapter, as well as Chapter 13, will be confined to a brief explanation of its principles and a discussion of its applications in interferometry.

12.1. The in-line (Gabor) hologram

In Gabor's original experiment, a transparency containing a small diffracting object on a clear background was illuminated with a collimated, monochromatic beam, and the interference pattern formed by the light diffracted by the transparency and the light transmitted through it (the coherent background) was recorded on a photographic plate. When a positive transparency made from this recording was illuminated with a collimated beam, it produced two diffracted waves, one of which was an exact reconstruction of the original object wave, while the other was its complex conjugate.

Gabor's experiment served to demonstrate the feasibility of holography, but the technique found little practical use for many years. This was mainly because of the poor quality of the reconstructed image, which was superposed on the conjugate image, as well as on a background of scattered light from the direct beam. This problem was finally solved by the use of a separate off-axis reference beam derived from the same source [Leith and Upatnieks, 1962, 1963, 1964].

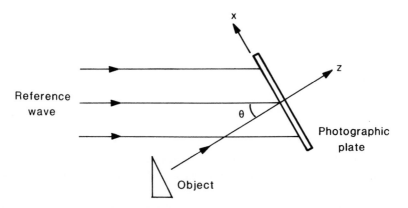

Fig. 12.1. Hologram recording with an off-axis reference beam.

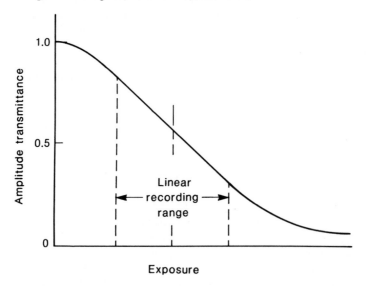

Fig. 12.2. Amplitude transmittance versus exposure curve for a typical hologram recording material.

12.2. The off-axis (Leith–Upatnieks) hologram

For simplicity, we shall assume that, as shown in Fig. 12.1, the reference beam is a collimated beam of uniform intensity incident at an angle θ on the photographic plate. The complex amplitude due to the reference beam at any point $P(x,y)$ on the plate can then be written as

$$r(x,y) = r \exp(i2\pi\xi_r x) \tag{12.1}$$

where $\xi_r = (\sin\theta)/\lambda$, while that due to the object beam is

$$o(x,y) = |o(x,y)| \exp[-i\phi(x,y)] \tag{12.2}$$

The resultant intensity in the interference pattern formed by these two waves is, therefore,

$$\begin{aligned}
I(x,y) &= |r(x,y) + o(x,y)|^2 \\
&= r^2 + |o(x,y)|^2 \\
&\quad + r|o(x,y)| \exp\{-i[2\pi\xi_r x + \phi(x,y)]\} \\
&\quad + r|o(x,y)| \exp\{i[2\pi\xi_r x + \phi(x,y)]\} \\
&= r^2 + |o(x,y)|^2 + 2r|o(x,y)| \cos[2\pi\xi_r x + \phi(x,y)]
\end{aligned} \tag{12.3}$$

Equation (12.3) shows that the intensity distribution consists of a set of fine fringes, constituting a carrier with a spatial frequency ξ_r, whose visibility and spacing are modulated by the amplitude and phase of the object wave.

If, as shown in Fig. 12.2, the recording material has a linear response to exposure, its amplitude transmittance can be written as

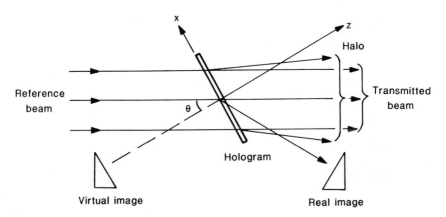

Fig. 12.3. Image reconstruction by a hologram recorded with an off-axis reference beam.

$$t = t_0 + \beta TI \qquad (12.4)$$

where t_0 is a constant corresponding to the transmittance of the unexposed material, T is the exposure time and β is the slope (negative) of the amplitude transmittance versus exposure characteristic of the material. The amplitude transmittance of the hologram is then

$$
\begin{aligned}
t(x,y) = t_0 &+ \beta T\Big[r^2 + |o(x,y)|^2 \\
&+ r|o(x,y)| \exp\{-\mathrm{i}[2\pi\xi_r x + \phi(x,y)]\} \\
&+ r|o(x,y)| \exp\{\mathrm{i}[2\pi\xi_r x + \phi(x,y)]\} \Big]
\end{aligned} \qquad (12.5)
$$

To reconstruct the image, the hologram is illuminated, as shown in Fig. 12.3, with the reference beam. The complex amplitude of the transmitted wave is then

$$
\begin{aligned}
u(x,y) &= r(x,y)\, t(x,y) \\
&= (t_0 + \beta T r^2)r \exp(\mathrm{i}2\pi\xi_r x) \\
&+ \beta T r |o(x,y)|^2 \exp(\mathrm{i}2\pi\xi_r x) \\
&+ \beta T r^2 o(x,y) \\
&+ \beta T r^2 o^*(x,y) \exp(\mathrm{i}4\pi\xi_r x)
\end{aligned} \qquad (12.6)
$$

Inspection of Eq. (12.6) shows that the first term on the right-hand side is merely the directly transmitted reference beam, while the second term corresponds to a halo around it. The third term, which is the one of interest, is the same as the original object wave, except for a constant factor, and generates a virtual image of the object in its original position. The fourth term corresponds to a conjugate wave which is propagated at an angle with the axis which is very nearly twice that made by the reference beam with it, and produces a real image.

An interesting result is that in this case, unlike the Gabor hologram, a 'positive' image is always obtained, even if the hologram recording is a photographic negative. In addition, if the angle between the object beam and the reference beam is made large enough, it is possible to view the virtual image without interference from the other transmitted and diffracted beams. The minimum value of θ for this is determined by the minimum spatial carrier frequency (ξ_r) for which the angular spectrum of the third term on the right hand side of Eq. (12.6) will not overlap the angular spectra of the other terms. Since these angular spectra are the Fourier transforms of the terms, there will be no overlap if

$$\xi_r \geqslant 3\xi_m \tag{12.7}$$

where ξ_m is the highest spatial frequency in the object beam.

12.2.1. Orthoscopic and pseudoscopic images

The virtual image formed by the hologram is located in the same position with respect to it as the object, and exhibits normal visual parallax. On the other hand, it can be seen from Fig. 12.3 that the real image has the curious property that its depth is inverted. Such an image is called a pseudoscopic image, as opposed to a normal or orthoscopic image [Leith and Upatnieks, 1964; Rosen, 1967]. This depth inversion results in conflicting visual clues which make viewing of the real image psychologically unsatisfactory. If required, an orthoscopic real image of an object can be produced, either by recording two holograms in succession [Rotz and Friesem, 1966], or by recording a hologram of an orthoscopic real image of the object formed by a large lens or a concave mirror.

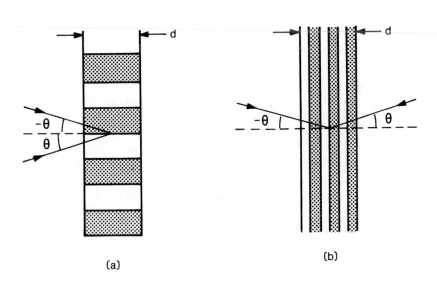

(a)   (b)

Fig. 12.4. Formation of (a) volume transmission, and (b) volume reflection holograms.

12.3. Volume holograms

We have assumed so far that the hologram recording can be treated as a two-dimensional grating. This assumption is valid only when the thickness of the recording medium is small compared to the average spacing of the hologram fringes. Typically, the spacing of the hologram fringes may be about 1 μm, while a photographic emulsion layer may have a thickness of 5 – 15 μm, and other recording materials may be much thicker. Under these conditions, the hologram has to be considered as a volume recording.

12.3.1. Volume transmission holograms

Consider the simplest case of a plane object wave and a plane reference wave incident, as shown in Fig. 12.4(a), at angles $\pm\ \theta$ on a recording material of thickness d. The hologram then consists of a set of planes running perpendicular to the surface of the recording material, with a spacing Λ given by the relation

$$2\Lambda \sin \theta = \lambda \tag{12.8}$$

It is then possible to define a parameter

$$Q = 2\pi\lambda d / n\Lambda^2 \tag{12.9}$$

where n is the average refractive index of the medium. Over a fairly wide range of variables, holograms for which $Q < 1$ can be treated as thin holograms, while those for which $Q > 1$ should be treated as volume holograms.

The characteristic feature of diffraction at a volume hologram is angular and wavelength selectivity. At the reconstruction stage, the diffracted amplitude is a maximum only when the angle of incidence θ_r and the wavelength λ_r of the beam used to illuminate the hologram satisfy the Bragg condition

$$\sin \theta_r = \pm \lambda_r / 2\Lambda \tag{12.10}$$

If $\lambda_r = \lambda$, $\theta_r = \pm\ \theta$. The Bragg condition is therefore satisfied when the hologram is illuminated by a duplicate of either the reference wave or the object wave. Away from the Bragg angle, the diffraction efficiency drops off rapidly.

12.3.2. Volume reflection holograms

If a hologram is recorded with the object and reference waves incident, as shown in Fig. 12.4(b), at angles of $\pm\ \theta$ from opposite sides of a thick recording material, the hologram consists of planes running parallel to the surface of the recording medium. This type of hologram, which was first proposed by Denisyuk [1962], has such a high wavelength selectivity that when it is illuminated with white light it reflects a narrow enough band of wavelengths to form a monochromatic image. The mechanism of wavelength selection in this case resembles very closely that in Lippmann's technique of colour photography (see Section 2.14).

12.4. Phase holograms

The amplitude transmittance of any recording medium is a complex quantity which can be written as

$$t = \exp(-\alpha d)\exp(-i2\pi nd/\lambda) \qquad (12.11)$$

where α is the absorption constant of the material, d is its thickness and n is its refractive index. If, as in the case of photographic materials, α changes with exposure to light, the resulting hologram is called an amplitude hologram. On the other hand, if $\alpha = 0$, and either d or n changes with exposure, the resulting hologram is called a phase hologram.

TABLE 12.1

Diffraction efficiency of gratings

Grating type	Variable		Maximum diffraction efficiency (%)
Thin transmission	{	Absorption	6.25
		Thickness	33.9
Volume transmission	{	Absorption	3.7
		Refractive index	100
Volume reflection	{	Absorption	7.2
		Refractive index	100

12.5. Diffraction efficiency

The diffraction efficiency of a hologram can be defined as the ratio of the power diffracted into the desired image to that incident on the hologram. A detailed theoretical analysis of the diffraction efficiency that can be obtained with holograms of different types recorded with two plane waves has been made by Kogelnik [1967, 1969], and his results are summarized in Table 12.1. As can be seen, a theoretical maximum diffraction efficiency of 100% is possible with a volume phase hologram. However, the diffraction efficiencies obtained with holograms of diffusely reflecting objects are considerably lower because the intensity of the object wave exhibits strong local fluctuations due to the formation of a speckle pattern [Upatnieks and Leonard, 1970].

12.6. Optical systems and light sources

Holography is mostly carried out with cw lasers, usually either a He–Ne laser or an Ar⁺ laser with an intra-cavity étalon to ensure operation in a single longitudinal mode. A typical setup is shown in Fig. 12.5. To avoid vibrations, all the components are mounted on a rigid optical table supported by pneumatic isolation mounts. The laser beam is divided at a beam-splitter with a graded metal coating which permits controlling the ratio of the intensities in

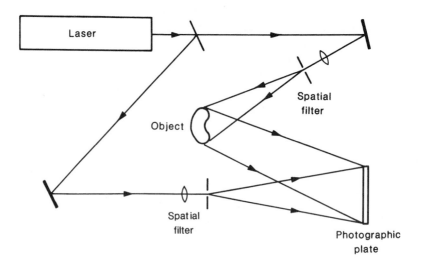

Fig. 12.5. Typical optical arrangement for recording a hologram.

the two beams, and two microscope objectives are used to expand the object and reference beams to the required size. A pinhole at the focus of each microscope objective serves as a spatial filter to eliminate unwanted diffracted light from dust and imperfections in the optics and give a clean beam.

12.7. Recording materials

A number of materials can be used for recording holograms [see Smith, 1977; and Hariharan, 1980, for a survey of their characteristics]. However, the most commonly used are high-resolution photographic materials, which have the advantages of high sensitivity, a wide range of spectral sensitivities to match available laser wavelengths and ready commercial availability. With normal processing they give amplitude holograms, but volume phase holograms with higher diffraction efficiency can be produced by using a bleach bath which replaces the silver image by one made up of a silver salt with a high refractive index. Processing normally results in a reduction in the thickness of a photographic emulsion layer because the unused silver halide is removed. In a volume reflection hologram, this causes a shift in the colour of the reconstructed image towards shorter wavelengths, which can be eliminated either by treating the hologram with a swelling agent or by using a tanning developer.

Two other materials which have been used fairly widely are photoresists and dichromated gelatin. Both of these have relatively low sensitivity, but give high quality recordings with low scattering. Photoresists give a surface relief image, while dichromated gelatin gives volume phase holograms with very high diffraction efficiency [Shankoff, 1968].

Photothermoplastics are also increasingly used because they can be processed in situ and recycled. This type of material consists of a substrate (glass or Mylar) coated first with a thin, transparent conducting layer of indium oxide, then with a photoconductor and, finally, with a thermoplastic [Urbach and Meier, 1966; Lin and Beauchamp, 1970]. The material is initially sensitized in darkness by spraying an electric charge uniformly on to the top surface. On exposure and recharging, a spatially varying charge pattern is produced. When the thermoplastic is heated briefly to its softening point, by passing a current through the indium oxide layer, the field due to this spatially varying charge distribution deforms the thermoplastic, resulting in a thin phase hologram with good diffraction efficiency. Finally, when the plate is to be re-used, it is flooded with light and the thermoplastic is heated once again to erase the hologram.

Electro-optic crystals are another promising class of materials for holography. The most interesting of these are photoconductive electro-optic crystals such as $Bi_{12}SiO_{20}$ (BSO) and $Bi_{12}GeO_{20}$ (BGO) [Huignard and Micheron, 1976]. A thin crystal slice is used, to which an electric field is applied by means of external electrodes. When the crystal is exposed to light, the stored field decays in the illuminated areas, and this results in a corresponding locally varying birefringence due to the Pockels effect. The recorded hologram can be erased by uniformly exposing the crystal, which can be recycled indefinitely. The only problem with these materials is that optical readout is destructive, because the same wavelength must be used for recording and readout to satisfy the Bragg condition. This problem can be overcome by storing the reconstructed image on a vidicon memory tube for subsequent observation.

12.8. Phase-conjugate imaging

An interesting application of holography is in the production of images unaffected by lens aberrations or by the presence of surfaces in the optical path which introduce aberrations [Kogelnik, 1965]. In this technique a hologram is recorded of the aberrated object wave using a collimated reference

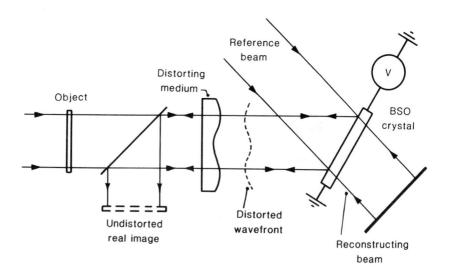

Fig. 12.6. Optical arrangement for phase-conjugate imaging in real time [Huignard et al., 1979].

beam of uniform intensity. If we assume linear recording, the amplitude transmittance of the hologram can be written from Eqs (12.3) and (12.4) as

$$t = t_0 + \beta T[|r|^2 + |o|^2 + r^*o + ro^*] \qquad (12.12)$$

where r and o are, respectively, the complex amplitudes of the reference and object waves at the hologram. If, then, this hologram is illuminated by the conjugate of the reference wave (that is to say, a wave equivalent to the reference wave travelling backwards in time) the complex amplitude of the wave transmitted by the hologram is

$$r^*t = r^*t_0 + \beta T[r^*|r|^2 + r^*|o|^2 + r^{*2}o + |r|^2o^*] \qquad (12.13)$$

Since the reference beam has a uniform intensity, $|r|^2$ is a constant and the last term within the square brackets corresponds to the reconstruction of the conjugate of the object wave. This has exactly the same phase errors as the original object wave, except that they have the opposite sign. Accordingly, when this wave propagates back through the optical system, the phase errors cancel out, so that an aberration-free real image is formed.

Phase-conjugate imaging can be implemented in real time with a recording medium such as BSO, using the optical system shown in Fig. 12.6 [Huignard et al., 1979], in which the collimated reference beam is reflected back through the BSO crystal, and the real image formed by the conjugate object wave is brought out by a beam-splitter. This technique has considerable potential in the production of very large scale integrated circuits where high-resolution imaging over a large field is necessary [Levenson, 1981]. As will be described in Chapter 13, it also has significant applications in interferometry.

12.9. Holographic optical elements

Conventional ruled diffraction gratings are now being replaced progressively by holographic diffraction gratings. These are produced by recording the interference pattern produced by two plane or spherical wavefronts in a thin layer of photoresist coated on an optically worked blank. The photoresist is processed to give a relief image, which is coated with an evaporated metal layer [Schmahl and Rudolph, 1976]. Holographic gratings have the advantage that they are free from both random and periodic errors and have low levels of scattered light; it is also possible to produce gratings with unique focusing properties. They can be blazed by a number of techniques [Hutley, 1976, 1982].

More general holographic optical elements (HOEs) can be used as beam-splitters or replacements for more complex optical components [see, for example, the review by Close, 1975]. The most commonly used material for HOEs is dichromated gelatin which can be processed to give very high diffraction efficiency [Chang and Leonard, 1979]. HOEs have the advantage that they can perform quite complex functions and, since they can be produced on thin substrates, are quite light, even for large apertures. They have been used very effectively in head-up displays and multiple-imaging systems as well as to replace mirror scanners in point-of-sale terminals. They also open up a number of interesting possibilities in interferometry, since, as will be shown in Section 12.10, they can be made to generate wavefronts with any prescribed amplitude and phase distribution.

12.10. Computer-generated holograms

Holograms can be synthesized by means of a digital computer. The production of such computer-generated holograms (CGH) has been discussed in detail by Lee [1978], Yaroslavskii and Merzlyakov [1980] and Dallas [1980] and involves two basic steps.

The first step is to calculate the complex amplitude of the object wave at the hologram plane. For convenience, this is usually taken to be the Fourier transform of the complex amplitude in the object plane. Since the transform has to be evaluated digitally, it is necessary to sample the wavefront in the object plane. If an image consisting of $N \times N$ resolvable elements is to be formed, it can be shown that the object wave must be sampled at $N \times N$ equally spaced points; the $N \times N$ coefficients of its discrete Fourier transform can then be calculated using the fast Fourier transform algorithm [Cochran et al., 1967].

The next step is to produce, using the calculated values of the discrete Fourier transform, a transparency (the hologram) which, when properly illuminated, reconstructs the object wave. This process can be considerably simplified if the hologram has only two levels of amplitude transmittance — zero and one. Such a hologram is called a binary hologram.

The best known hologram of this type is the binary detour-phase hologram [Brown and Lohmann, 1966, 1969]. In this, the area of the hologram is divided into $N \times N$ cells, and each complex coefficient of the discrete Fourier transform is represented by a single transparent window within the corresponding cell. If such a hologram is illuminated at an angle by a collimated beam, it is apparent that a shift of the window within any cell would result in the light diffracted by it travelling by a longer or shorter path to the reconstructed image. Accordingly, it is possible to encode the complex coefficients by making the areas of the windows proportional to the moduli of these coefficients, while their phases are represented by the positions of the windows within the cells. Figure 12.7(a) shows a typical binary detour-phase hologram, while Fig. 12.7(b) shows the image reconstructed by it. The first-order images are those just above and below the central spot.

a

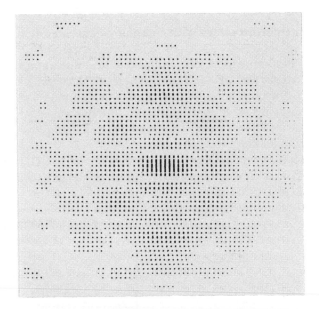

b

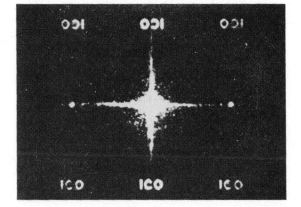

Fig. 12.7. (a) Binary detour-phase hologram, and (b) reconstructed image (Lohmann and Paris, 1967].

Binary detour-phase holograms have the advantage that a simple plotter can be used to prepare a master which can then be photographed on a reduced scale. However, to minimize noise due to quantization of the modulus and phase of the Fourier coefficients, the number of addressable points in each cell must be fairly large. A number of improved encoding techniques have been described to get around this problem [Lee, 1974, 1979].

12.11. Interferometry with computer-generated holograms

Normally, interferometric tests of aspheric optical surfaces require either an aspheric reference surface or an additional element, commonly called a null lens, which converts the wavefront leaving the surface under test into a spherical or plane wavefront. A simpler alternative is to use a CGH; [MacGovern and Wyant, 1971]; this technique is now employed quite widely [see, for example, the review by Loomis, 1980].

A modified Twyman–Green interferometer using a CGH to test an aspherical mirror is shown in Fig. 12.8 [Wyant and Bennett, 1972]. The CGH is located in the plane in which the mirror under test is imaged and is equivalent to the interferogram formed by the wavefront from an aspheric

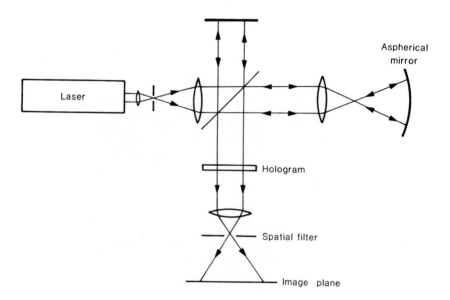

Fig. 12.8. Modified Twyman–Green interferometer using a computer-generated hologram to test an aspherical mirror [Wyant and Bennett, 1972].

surface with the desired profile and a tilted plane wavefront. The moiré pattern formed by the superposition of the actual interference fringes and the CGH gives the deviation of the surface under test from the ideal surface. The contrast of the moiré pattern is improved by re-imaging the hologram. A small aperture placed in the focal plane of the imaging lens passes only the wavefront from the mirror under test which has been transmitted by the hologram and the wavefront reconstructed by the hologram when it is illuminated with a plane wavefront. Typical fringe patterns obtained with such a setup, without and with the CGH, are shown in Fig. 12.9.

A limit is set to the deviations from a sphere which can be handled by a CGH by the requirement that the spatial carrier frequency of the CGH must be at least three times the highest spatial frequency in the uncorrected interference pattern. This restriction can be overcome by using a combination of a simple null lens, which reduces the residual aberrations to an acceptable level, and a CGH [Faulde et al., 1973; Wyant and O'Neill, 1974].

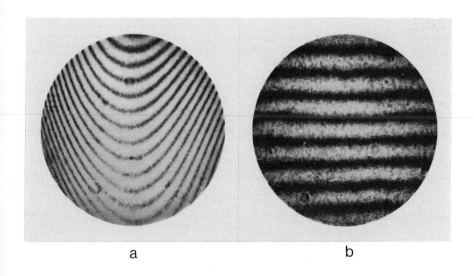

a   b

Fig. 12.9. Interferograms of an aspherical surface (a) without, and (b) with, a compensating computer-generated hologram [Wyant and Bennett, 1972].

13
Holographic interferometry

In holographic interferometry, one of the interfering wavefronts is stored in a hologram and then compared with another wavefront. Alternatively, interference can take place between two stored wavefronts. This makes it possible to compare wavefronts which were originally separated in time or space, or even wavefronts originally corresponding to different wavelengths.

Perhaps the most important advantage of holographic interferometry is that changes in the shape of objects with quite rough surfaces can be measured with very high accuracy. As a result, holographic interferometry has been used extensively in nondestructive testing, biomedical engineering and stress analysis. Techniques and applications in these fields have been reviewed by Erf [1974], Greguss [1975], von Bally [1979] and Vest [1979].

13.1. Real-time holographic interferometry

When a hologram is replaced in its original position in the setup in which it was recorded, it reconstructs the original object wave. If, then, the shape of the object changes slightly, interference between the reconstructed object wave, which corresponds to the undeformed object, and the directly transmitted object wave, gives rise to a pattern of fringes covering the object, which can be used to map the changes in shape of the object in real time.

In this case, since the hologram is illuminated by the object beam as well as the reference beam, the complex amplitude of the wave transmitted by the hologram is

$$u(x,y) = [r(x,y) + o'(x,y)]t(x,y) \qquad (13.1)$$

where $r(x,y)$ is the complex amplitude of the reference wave, $o'(x,y)$ is the complex amplitude of the wave from the deformed object and $t(x,y)$ is the amplitude transmittance of the hologram. If the change in the shape of the object is small, only the phase distribution of the object wave is modified, so that

$$o'(x,y) = o(x,y) \exp[-i\Delta\phi(x,y)] \qquad (13.2)$$

where $o(x,y)$ is the complex amplitude of the original object wave and $\Delta\phi(x,y)$ is the change in phase.

From Eqs (12.5) and (13.1), the complex amplitude of the directly transmitted object wave is

$$u_1(x,y) = (t_0 + \beta Tr^2)o'(x,y) \qquad (13.3)$$

where $r^2 = |r(x,y)|^2$. Similarly, the complex amplitude of the reconstructed object wave is

$$u_2(x,y) = \beta Tr^2 o(x,y) \qquad (13.4)$$

Accordingly, the intensity in the interference pattern is

$$I(x,y) = |u_1(x,y) + u_2(x,y)|^2$$
$$= |o(x,y)|^2 [A^2 + (t_0-A)^2 - 2A(t_0-A) \cos \Delta\phi(x,y)] \qquad (13.5)$$

where $A = -\beta Tr^2$. Since β is negative, a dark fringe corresponds to the condition

$$\Delta\phi(x,y) = 2m\pi \qquad (13.6)$$

where m is an integer.

The problem of replacing the hologram in exactly the same position can be solved by processing it *in situ* in a liquid gate; this can be done conveniently with a monobath [Hariharan and Ramprasad, 1973b]. Alternatively, a photothermoplastic can be used as the recording material.

13.2. Double-exposure holographic interferometry

In this technique, two holograms are recorded on the same photographic plate, one of the object in its initial condition, and the other of the deformed object. If we assume linear recording, the hologram will reconstruct these two waves, so that the resultant complex amplitude, apart from a constant of proportionality, is

$$u(x,y) = o(x,y) + o'(x,y)$$
$$= o(x,y) \{1 + \exp[-i\Delta\phi(x,y)]\} \tag{13.7}$$

The intensity in the image is then

$$I(x,y) = |u(x,y)|^2$$
$$= |o(x,y)|^2 [1 + \cos \Delta\phi(x,y)] \tag{13.8}$$

In this case, bright fringes are obtained when

$$\Delta\phi(x,y) = 2m\pi \tag{13.9}$$

Double-exposure hologram interferometry has the advantage that the two interfering waves are reconstructed in exact register, and no special care need be taken in illuminating the hologram. The visibility of the fringes is also good, since the two waves always have the same polarization and very nearly the same amplitude. The hologram is a permanent record of the changes in the shape of the object between the first and second exposures.

13.3. Holographic interferometry of phase objects

An important area of application of holographic interferometry is in studies of transparent phase objects. Even though conventional interferometry can be used for this purpose, holographic interferometry has the advantage that the optical quality of the mirrors and windows is not critical, since the phase errors contributed by them affect both wavefronts equally and cancel out. In addition, if the object is illuminated through a diffuser, the interference pattern can be viewed over a range of angles, making it possible to study three-dimensional refractive index distributions [Sweeney and Vest, 1973; Cha and Vest, 1979, 1981].

Techniques such as multiple-beam and shearing interferometry can also be applied in holographic interferometry [see Vest, 1979]. An interesting type of shear which is possible only with holography is longitudinally-reversed shear (phase-conjugate interferometry), in which the primary and conjugate images of a wavefront reconstructed by a hologram are made to interfere [Bryngdahl, 1969]. Optical systems which permit phase-conjugate interferometry in real time have been described by Bar-Joseph *et al.* [1981] and Ja [1982]. Phase difference amplification is also possible using the higher diffracted orders from a nonlinear hologram [Bryngdahl and Lohmann, 1968].

A useful application of holographic interferometry is in studies of plasmas. For this, two holograms are recorded simultaneously on the same plate with light from a ruby laser which has passed through a frequency-doubling crystal to produce two collinear beams with wavelengths of 694 nm and 347 nm. If the second-order image reconstructed by one hologram is made to interfere with the first-order image reconstructed by the other, interference fringes are obtained which are contours of constant dispersion and, hence, of constant electron density [Ostrovskaya and Ostrovskii, 1971].

13.4. Interferometry with diffusely reflecting objects

Holographic interferometry makes it possible to obtain interference patterns having good visibility with an object having a rough surface, since both the

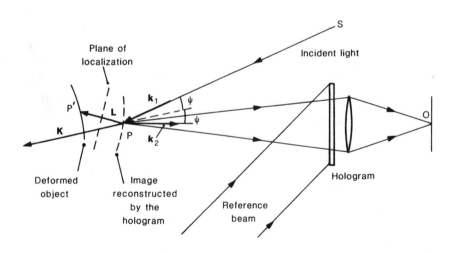

Fig. 13.1. Holographic interferometry with a diffusely reflecting object.

interfering wavefronts exhibit exactly the same microstructure. The average intensity at any point and, hence, the interference order, is determined by the phase difference between corresponding points on the two wavefronts. To evaluate this phase difference, we consider a small area of the surface which has been subjected to a simple translation, as shown in Fig. 13.1, so that a point P undergoes a vector displacement L to P'. If the displacement of P is small compared to the optical paths to the source S and the observer O, the phase difference

$$\Delta\phi = L \cdot (k_1 - k_2)$$
$$= L \cdot K \qquad (13.10)$$

where k_1 and k_2 are the propagation vectors of the incident and scattered light, and $K = k_1 - k_2$ is the sensitivity vector [Aleksandrov and Bonch-Bruevich, 1967; Ennos, 1968; Sollid, 1969]. The vectors k_1 and k_2 are drawn along the directions of illumination and observation, respectively, and their magnitude is $|k_1| = |k_2| = 2\pi/\lambda$, while the sensitivity vector lies along the bisector of the angle 2ψ between the directions of illumination and viewing, and its magnitude is $|K| = (4\pi/\lambda) \cos\psi$.

An early observation was that the visibility of the interference fringes was a maximum for a particular position of the plane of observation, and that this position depended on the type of displacement undergone by the object. This phenomenon of localization of the fringes has obvious similarities to that observed in conventional interferometry with an extended source (see Section 2.5), and has been studied in detail by a number of authors [Stetson, 1969, 1970a, 1974; Walles, 1969; Steel, 1970, and Dubas and Schumann, 1974].

A simple treatment is possible based on the fact that only waves from corresponding points on the two wavefronts contribute to the fringes. If we consider, as in Fig. 13.1, the phase difference $\Delta\phi$ at an arbitrary plane between the waves from such a pair of corresponding points (P and P'), we find that this phase difference will, in general, vary over the range of viewing directions defined by the aperture of the viewing system due to changes in the sensitivity vector. There will be, however, over a limited region, a position of this plane at which the value of $\Delta\phi$ is very nearly constant over this range of viewing directions, and this position corresponds to the plane of localization.

Two special cases of interest are those of pure translation of the surface and pure rotation about an axis in the surface. It can be shown that, in both cases, the fringes are parallel straight lines; however, in the former case they are localized at infinity while in the latter case they are localized very close to the surface.

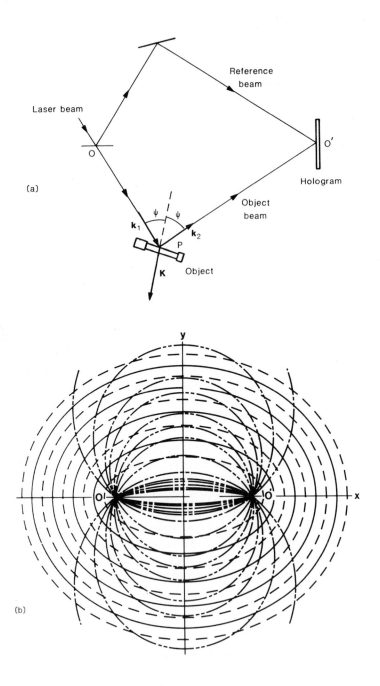

Fig. 13.2.   (a) A simplified holographic setup, and (b) the holodiagram [Abramson, 1969].

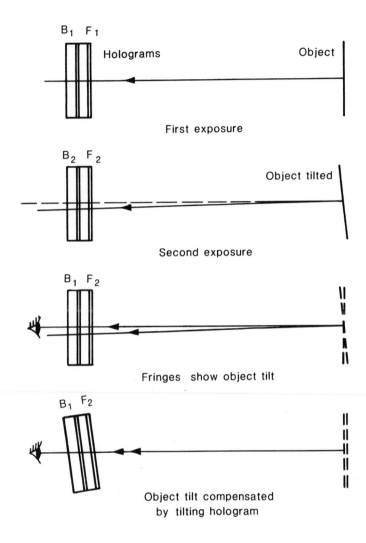

Fig. 13.3. The sandwich hologram [Abramson, 1975].

13.5. The holodiagram

A simple geometrical representation of these relations can be obtained from the holodiagram [Abramson, 1969b, 1970a,b, 1971, 1972]. For a simple holo-gram recording system such as that shown in Fig. 13.2(a), the holodiagram consists, as shown in Fig. 13.2(b), of a set of ellipses with their foci at O and

O', each of which is the locus of points for which the distance OPO' is a constant. This distance changes by one wavelength from one ellipse to the next, corresponding to a change in the interference order of unity. The holodiagram shows that the displacement of P required for this change is a minimum when its motion is normal to the ellipse, that is to say, along the sensitivity vector $\boldsymbol{K}$. The circles passing through O and O' are the loci of points for which the angle $OPO' = 2\psi$ is a constant, and correspond, there-fore, to constant values of the sensitivity vector.

The holodiagram can be used to design a holographic system to max-imize or minimize its sensitivity to a particular type of surface displacement, as well as to interpret the fringe patterns obtained for different types of displacements. It can also be used to minimize the variations in the optical path over a large object, when recording a hologram of it.

13.6. Sandwich holograms

Double-exposure holography normally suffers from several limitations. Thus, it is only possible to compare two states of the object. In addition, it is difficult to separate rigid body displacements and deformations of the object or to identify unambiguously the sign of the surface displacements. However, these limitations can be overcome by a number of techniques.

One way of determining the sense of a displacement is to introduce a small tilt in the object wavefront between the two exposures; this produces a system of reference fringes whose position is modulated by the phase shifts being studied [Hariharan and Ramprasad, 1973a]. Multiplexing techniques using a set of overlapping masks can be used to record a sequence of holograms to study the deformation of the object at different stages of loading [Hariharan and Hegedus, 1973]. Control of the fringes to compen-sate for rigid body motion is possible with two holograms recorded with two different, angularly separated reference waves [Gates, 1968; Ballard, 1968; Tsuruta et al., 1968].

An elegant alternative is the sandwich hologram [Abramson, 1974, 1975, 1977; Abramson and Bjelkhagen, 1979]. In this technique, as shown in Fig. 13.3, pairs of photographic plates are exposed simultaneously with their emulsion-coated surfaces facing the object. Thus, B_1, F_1 are exposed with the object in its original state, while B_2, F_2, and B_3, F_3, ... are exposed as the object is progressively deformed. If, finally, B_1 is combined with F_2, F_3, . . . , the total deformation at any stage can be studied, while combinations such as $B_1 F_2$, $B_2 F_3$, $B_3 F_4$, . . . , show the incremental deformations. A tilt of the sandwich acts on the fringes in the same way as a tilt of the object. A simpler version of this technique, using only two plates cemented together with a spacer, has been described by Hariharan and Hegedus [1976], while a theoretical analysis of the changes in the fringes has been made by Dubas and Schumann [1977].

13.7. Holographic strain analysis

Quantitative stress analysis requires measurement of the strains. Several methods of analysis of the fringe pattern have been described for this purpose, and most of these have been discussed in an exhaustive review by Briers [1976].

A direct method which is of considerable interest is the fringe vector method [Stetson, 1974, 1975a,b, 1979]. This makes use of the fact that any combination of homogeneous deformation and rotation of an object gives rise to fringes corresponding to the intersection of the object surface with a series of equally spaced surfaces of constant phase difference. Experimental studies using this method have been described by Pryputniewicz and Stetson [1976, 1980], Pryputniewicz and Bowley [1978] and Pryputniewicz [1978, 1980].

The other approach which has been followed is to measure the actual surface displacements and differentiate them. For this, it is necessary to make three observations of the fringe order from at least three different directions [Shibayama and Uchiyama, 1971] or, alternatively, to use a single direction of observation and three, or better, four directions of viewing [Hung et al., 1973; Goldberg, 1975]. To obtain adequate accuracy, television techniques have been used to view and process the fringes [Nakadate et al., 1981]; a better approach is to use electronic techniques such as heterodyne interferometry and digital interferometry, which permit direct measurements of the optical phase difference at a uniformly spaced network of points [Dändliker et al., 1973; Dändliker, 1980; Hariharan et al., 1983].

13.8. Holographic interferometry with pulsed lasers

Double-exposure holograms made with a pulsed ruby laser make it possible to study transient phenomena as well as to carry out measurements in an industrial environment. [Gates et al., 1972; Armstrong and Forman, 1977]. Objects rotating at high speeds can also be studied by means of an optical derotator [Stetson, 1978]. This consists of an inverting prism which rotates at half the speed of the object, giving an image which is almost stationary.

13.9. Time-average holographic interferometry

Time-average holographic interferometry is a technique which can be used to map the amplitude of vibration of a diffusely reflecting surface [Powell and Stetson, 1965]. For this, a hologram is recorded with an exposure time which is much longer than the period of vibration.

Let $o(x,y) = |o(x,y)| \exp[-i\phi(x,y)]$ represent the complex amplitude of the light scattered by a point $P(x,y)$ on the object when it is at rest. If, then, the displacement of P at time t is given by the relation

$$L(x,y,t) = L(x,y) \sin \omega t \tag{13.11}$$

the complex amplitude of the light becomes

$$o(x,y,t) = |o(x,y)| \exp \{ -i[\phi(x,y) + K \cdot L(x,y) \sin \omega t] \} \tag{13.12}$$

where K is the sensitivity vector.

We can think of the hologram, in this case, as the superposition of a large number of recordings for slightly different positions of the object. If this process is linear, the complex amplitude $u(x,y)$ of the reconstructed wave is proportional to the time average of $o(x,y,t)$, so that we have

$$
\begin{aligned}
u(x,y) &= (1/T) \int_0^T |o(x,y)| \exp \{ -i[\phi(x,y) + K \cdot L(x,y) \sin \omega t] \} \, dt \\
&= |o(x,y)| \exp[-i\phi(x,y)](1/T) \int_0^T \exp[-iK \cdot L(x,y) \sin \omega t] dt \\
&= |o(x,y)| M_T(x,y)
\end{aligned}
\tag{13.13}
$$

where T is the exposure time, and $M_T(x,y)$ is known as the characteristic function.

Since $T \gg 2\pi/\omega$, we can write

$$
\begin{aligned}
M_T(x,y) &= \lim_{T \to \infty} (1/T) \int_0^T \exp[-iK \cdot L(x,y) \sin \omega t] dt \\
&= J_0[K \cdot L(x,y)]
\end{aligned}
\tag{13.14}
$$

where J_0 is the zero-order Bessel function of the first kind. Accordingly, the intensity in the reconstructed image is

$$
\begin{aligned}
I(x,y) &= |o(x,y) M_T(x,y)|^2 \\
&= I_0(x,y) J_0^2(\Omega)
\end{aligned}
\tag{13.15}
$$

where $I_0(x,y)$ is the intensity distribution for the stationary object, and $\Omega =$ $\boldsymbol{K} \cdot \boldsymbol{L}(x,y)$.

Variations of the vibration amplitude across the object give rise to fringes (contours of equal vibration amplitude). The dark fringes correspond to the zeros of the function $J_0^2(\Omega)$, which is plotted in Fig. 13.4 (the solid line), and the bright fringes to its maxima. A typical series of interferograms showing the time-average fringes obtained with a model of an aircraft tail-fin excited at different frequencies is presented in Fig.13.5 [Abramson and Bjelkhagen, 1973]. These permit ready identification of the vibration modes and measurements of the vibration amplitude at any point.

13.10. Real-time interferometry of vibrating surfaces

In this technique, a hologram is recorded of the stationary object, and the reconstructed wave interferes with the wave from the vibrating object [Stetson and Powell, 1965]. If we assume that the reconstructed wave and the transmitted object wave have the same intensities, the resultant intensity at any instant t is, from Eq. (13.5),

$$I(x,y,t) = I_0(x,y)\{1 - \cos[\boldsymbol{K} \cdot \boldsymbol{L}(x,y) \sin \omega t]\} \tag{13.16}$$

However, since the time of response of the human eye (40 ms) is usually much greater than the period of vibration, the observer actually sees the time-averaged intensity

$$< I(x,y)> = I_0(x,y) \lim_{T \to \infty} (1/T) \int_0^T 1 - \cos[K \cdot L(x,y) \sin \omega t] \ dt$$

$$= I_0(x,y) [1 - J_0(\Omega)] \tag{13.17}$$

The function $[1 - J_0(\Omega)]$ is also plotted in Fig. 13.4 (the broken line), and corresponds to a dark field interferogram with half the number of fringes seen with the time-average technique.

Because of their relatively low contrast, real-time fringes are used mainly to identify the resonances of a test object, while varying the excitation frequency and the point of excitation, after which a time-average hologram can be made for more accurate measurements.

Very interesting results can be obtained by the use of real-time phase-conjugate imaging for the study of vibrating objects [Marrakchi, Huignard, and Herriau, 1980]. In this case, time-average fringes are formed because, to build up a hologram, the recording medium (BSO) (see Section 12.7) integrates over a time interval which is long compared to the period of the

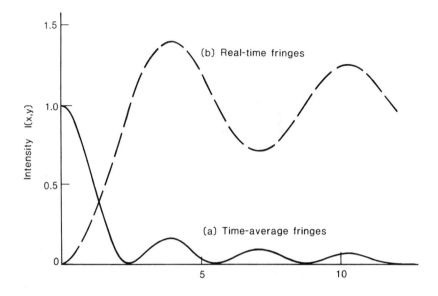

Fig. 13.4. Intensity distribution in the image of a vibrating object: (a) time-average fringes, (b) real-time fringes.

vibration. However, it is possible to observe, in real time, the changes in the time-average fringes as the excitation frequency is changed.

13.11 Stroboscopic holographic interferometry

If a hologram of a vibrating object is recorded using light pulses triggered at times t_1 and t_2 during the vibration cycle, the intensity in the image is

$$I(x,y) = I_0(x,y) \{1 + \cos[\boldsymbol{K} \cdot \boldsymbol{L}(x,y) (\sin \omega t_1 - \sin \omega t_2)]\} \quad (13.18)$$

The interference fringe pattern is equivalent to that given by a double-exposure hologram recorded with the object in these two states. The phase of the vibration can be determined if, say, t_2 can be varied, keeping t_1 fixed.

13.12. Holographic subtraction

This is a simple modification of the time-average technique, in which one exposure is made with the object stationary, followed by another with the

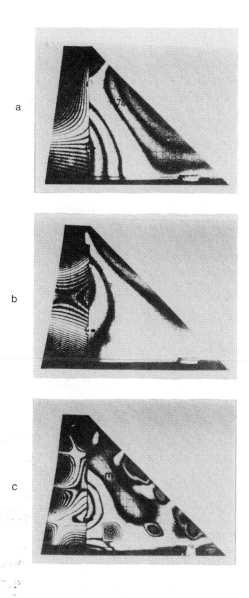

Fig. 13.5. Time-average interferograms of a model of an aircraft tail fin vibrating at (a) 670 Hz, (b) 894 Hz and (c) 3328 Hz [Abramson and Bjelkhagen, 1973].

object vibrating and with a phase shift of π in one of the beams [Hariharan, 1973]. The intensity distribution in the image is then

$$I(x,y) = I_0(x,y) [1 - J_0(\Omega)]^2 \qquad (13.19)$$

The characteristic function is the square of that for real-time holography. Since a perfectly dark field can be obtained, the technique gives a useful increase in sensitivity for small vibration amplitudes.

13.13. Temporally modulated holography

Holography with a temporally modulated reference beam is a very powerful method for the study of vibrations [Aleksoff, 1971]. The techniques which have been studied include frequency translation [Zambuto and Fischer, 1973], amplitude modulation [Takai et al., 1976] and phase modulation [Neumann et al., 1970]. These techniques make it possible to increase or decrease the sensitivity as well as to determine the variation of the relative phase across a vibrating surface.

13.14. Holographic contouring

Holographic interferometry can be used to generate contours of constant elevation with respect to a reference plane. Basically, the technique involves producing two images of the object, one of which has its scale changed appropriately along a specified direction, so that the interference fringes produced are the desired contours.

13.14.1. Two-wavelength contouring

In the two-wavelength contouring technique, a telecentric system images the object on the hologram as shown in Fig. 13.6, [Haines and Hildebrand, 1965; Hildebrand and Haines, 1966, 1967; Zelenka and Varner, 1968]. A plane wave is used to illuminate the object and another plane wave is used as the reference. Two exposures are made with different wavelengths, λ_1 and λ_2.

When the hologram is illuminated with one of the wavelengths, say λ_2, it reconstructs two images with the same lateral magnification (unity). However, the two reconstructed images of a point in the object are separated in the longitudinal direction by a distance

$$\Delta z = z(\lambda_1 - \lambda_2)/\lambda_1 \qquad (13.20)$$

where z is the original distance of this point from the surface of the photographic plate. Successive fringes then correspond to an increment δz of Δz equal to $\lambda_2/2$, so that

$$\delta z = \lambda_1\lambda_2/2(\lambda_1 - \lambda_2) \qquad (13.21)$$

Typically, with an Ar^+ laser, the two lines at $\lambda = 514$ nm and 488 nm give contours at an interval of approximately 5 μm. With a dye laser, the contour interval can be varied continuously [Friesem and Levy, 1976].

13.14.2. Two-refractive-index contouring

This method of contouring requires only a single wavelength. [Tsuruta et al., 1967; Zelenka and Varner, 1969]. As shown in Fig. 13.7, the object, which is placed in a cell with a plane glass window and imaged by a telecentric system, is illuminated with a plane wave by means of a beam-splitter. Two holograms are recorded on a plate placed near the stop of the telecentric system, with the cell filled with fluids having refractive indices n_1 and n_2,

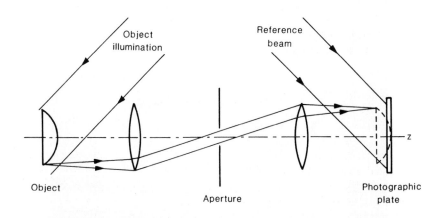

Fig. 13.6. Optical system for holographic contouring using two wavelengths [Zelenka and Varner, 1968].

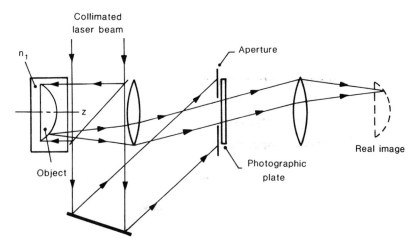

Fig. 13.7. Optical system for holographic contouring using the two-index technique [Zelenka and Varner, 1969].

Fig. 13.8. Contoured image of a coin obtained by the two-index method (contour interval 45 μm).

respectively. In this case, the two images are longitudinally displaced with respect to each other by an amount

$$\Delta z = (n_1 - n_2)z \qquad (13.22)$$

where z is the distance from the surface of the object to the window, so that fringes are obtained corresponding to increments of z given by the relation

$$|\delta z| = \lambda/2(n_1 - n_2) \qquad (13.23)$$

A very wide range of contour intervals (1 μm to > 300 μm) can be obtained by using combinations of liquids and gases.

Figure 13.8 shows typical contours obtained with a coin using the two-refractive-index technique.

14
Speckle interferometry

When the scattered light from a diffuser illuminated by a coherent source such as a laser falls on a screen, a stationary granular pattern results, called a speckle pattern (see Fig. 14.1). This phenomenon was initially regarded as a nuisance, since the image of a diffusing object formed with laser light is always covered with speckle. However, it was soon realized that speckle could be regarded as a random spatial carrier on which information on the shape and position of the diffusing surface was encoded. As a result, speckle quickly found a number of applications [see, for example, Dainty, 1975; Ennos, 1978; Erf, 1978]. This chapter summarizes some of the characteristics of laser speckle and describes how it is used in the techniques of speckle photography and speckle interferometry.

14.1. Speckle statistics

Speckle has its origin in the fact that most surfaces are extremely rough on a scale of light wavelengths. When such a rough surface is illuminated by a coherent source, the complex amplitude at a point $P(x,y)$ in the observation plane is the sum of the complex amplitudes of the diffracted waves from N microscopic elements making up the surface and can be written as

$$A(x,y) = N^{-\frac{1}{2}} \sum_{m=1}^{N} |A_m| \exp(-i\phi_m) \qquad (14.1)$$

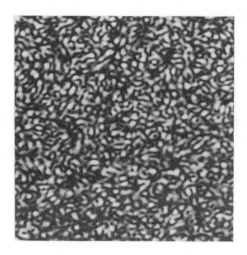

Fig. 14.1. Speckle pattern formed by scattered light from a diffuser illuminated by a laser beam.

We can assume that the amplitude and the phase of any diffracted component are statistically independent of each other, and are also statistically independent of the amplitudes and phases of all the other diffracted components. In addition, we will also assume that the differences in the optical paths from the point of observation to different elements cover a range of several wavelengths, so that the phase shifts ϕ_m, after subtracting integral multiples of 2π, are uniformly distributed over the interval 0 to 2π. Under these conditions, the real and imaginary parts of the complex amplitude obey Gaussian statistics, and their joint probability density function is (Goodman, 1975; Dainty, 1976]

$$P_{r,i}\,(A^{(r)},A^{(i)}) = \frac{1}{2\pi\sigma^2}\,\exp\left\{-\frac{[A^{(r)}]^2 + [A^{(i)}]^2}{2\sigma^2}\right\} \qquad (14.2)$$

where

$$\sigma^2 = \lim_{N \to \infty} (1/2N) \sum_{m=1}^{N} |A_m|^2 \qquad (14.3)$$

The mean value of the resultant amplitude $A(x,y)$ is zero, while the resultant phase has a uniform circular distribution.

Equation (14.2) can be transformed to give the probability density function of the intensity, which is

$$P_I(I) = (1/2\sigma^2) \exp(-I/2\sigma^2) \tag{14.4}$$

Equation (14.4) shows that the intensity in the speckle pattern has a negative exponential distribution. This is plotted in Fig. 14.2 (the solid line), from which it is apparent that the most probable intensity is zero.

It also follows from Eq. (14.4) that the mean value of the intensity is

$$<I> = 2\sigma^2 \tag{14.5}$$

while its second moment is

$$<I^2> = 2<I>^2 \tag{14.6}$$

The variance of the intensity is, therefore.

$$\begin{aligned} \sigma_I^2 &= <I^2> - <I>^2 \\ &= <I>^2 \end{aligned} \tag{14.7}$$

If we define the contrast of the speckle pattern as the ratio

$$C = \sigma_I/<I> \tag{14.8}$$

we can see, from Eq. (14.7), that the contrast of a speckle pattern formed with coherent light is unity.

14.2. Second-order statistics of speckle patterns

The second-order statistics of the intensity distribution in the speckle pattern can be evaluated [Goldfischer, 1965; Goodman, 1975] by making use of the fact that the complex amplitude $A(x_1,y_1)$ at the scattering plane and the complex amplitude $A(x,y)$ at the observation plane are related by the Fresnel–Kirchhoff integral. It can then be shown that the autocorrelation function of the intensity in the speckle pattern is given by the relation

$$R_I(\Delta x, \Delta y) = \langle I \rangle^2 \left\{ 1 + \left| \frac{\int\int_{-\infty}^{\infty} |A(x_1,y_1)|^2 \exp[(i2\pi/\lambda z)(x_1\Delta x + y_1\Delta y)]dx_1 dy_1}{\int\int_{-\infty}^{\infty} |A(x_1,y_1)|^2 dx_1 dy_1} \right|^2 \right\}$$

$$\tag{14.9}$$

The average dimensions of a speckle can be calculated from the width of the autocorrelation peak. For a square scattering surface whose edges have a length L, they are

$$\Delta x = \Delta y = \lambda z / L \qquad (14.10)$$

The other quantity of interest is the power spectral density of the intensity distribution, which is given by the Fourier transform of the autocorrelation function. We have

$$S_I(s_x, s_y) = \langle I \rangle^2 \left\{ \delta(s_x, s_y) \right.$$
$$\left. + \frac{\iint\limits_{-\infty}^{\infty} |A(x_1, y_1)|^2 |A(x_1 - \lambda z s_x, y_1 - \lambda z s_y)|^2 \, dx_1 dy_1}{[\iint\limits_{-\infty}^{\infty} |A(x_1, y_1)|^2 \, dx_1 dy_1]^2} \right\} \qquad (14.11)$$

where s_x, s_y are spatial frequencies along the x and y axes, respectively. The power spectrum is made up of two components, each containing half the total power. One is a delta function at the origin ($s_x = s_y = 0$), while the other is the normalized autocorrelation function of the intensity distribution over the scattering surface.

14.3. Image speckle

The average dimensions of the speckles in the image of a scattering surface formed by a lens are determined by the aperture of the lens, which acts as a low-pass spatial filter [Lowenthal and Arsenault, 1970]. The situation, in this case, can be analysed quite simply by considering the entrance pupil of the optical system as being illuminated by the primary speckle pattern. This random field then appears in the exit pupil, so that the statistics of the final speckle pattern can be obtained by treating the exit pupil as a rough object. For a circular pupil of radius ρ, the average size of the speckles in the image is then, from Eq. (14.9),

$$\Delta x = \Delta y = 0.61 \lambda f / \rho \qquad (14.12)$$

where f is the focal length of the optical system.

14.4. Addition of speckle patterns

When two speckle patterns are superposed, two limiting cases can arise. If the light fields are coherent, the amplitudes add, and it can be shown that the first-order statistics of the pattern do not change. On the other hand, if the light fields are incoherent, the intensities add. The probability density function of the intensity in the pattern formed by superposing two speckle patterns with equal average intensities is then [Burch, 1972; Goodman, 1975]

$$p(I) = (4I/<I>^2) \exp(-2I/<I>) \qquad (14.13)$$

This function is also plotted in Fig. 14.2, (the broken line). As can be seen, the main difference from the probability density function for a single speckle pattern with the same average intensity is the elimination of most of the dark areas. As a result, the contrast of the speckle pattern, as defined by Eq. (14.8), falls to $2^{-\frac{1}{2}}$.

14.5. Effect of depolarization

A factor which has an important effect on the statistics of a speckle pattern is the degree of depolarization caused by scattering at the surface. If the light is depolarized, the resulting speckle field can be considered as the sum of two component speckle fields produced by scattered light polarized in two orthogonal directions, and the intensity at any point is the sum of the intensities

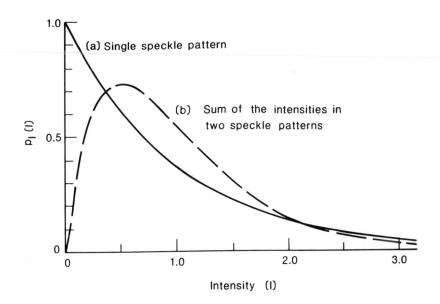

Fig. 14.2. Probability density function of the intensity: (a) a single speckle pattern with $<I> = 1$; (b) sum of the intensities in two speckle patterns with $<I_1> = <I_2> = 0.5$.

of these component speckle patterns [Goodman, 1975]. Since the component speckle patterns are only partially correlated, a decrease results in the contrast of the resultant speckle pattern, which depends on the illumination and viewing geometry [Hariharan, 1977]. To overcome this problem, a polarizer, which effectively transmits only one of the component speckle patterns, should be used in the viewing system.

14.6. Formation of Young's fringes

A case of some interest is when two identical speckle patterns are recorded on a photographic film with a small mutual displacement. If we assume linear recording as defined by Eq. (12.4), the resulting amplitude transmittance is

$$t(x,y) = t_0 + \beta T\,[I(x,y) + I(x,y-y_0)]$$
$$= t_0 + \beta TI(x,y) * [\delta(x,y) + \delta(x,y-y_0)] \qquad (14.14)$$

where $I(x,y)$ is the intensity distribution in the individual speckle patterns and y_0 is the amount by which they are displaced. If this transparency is illuminated, as shown in Fig. 14.3, by a collimated beam of monochromatic light, the amplitude in the back focal plane of the lens is proportional to the Fourier transform of $t(x,y)$, which is

$$g(u,v) = t_0\,\delta(u,v) + \beta Th(u,v)[1 + \exp{(ikvy_0)}] \qquad (14.15)$$

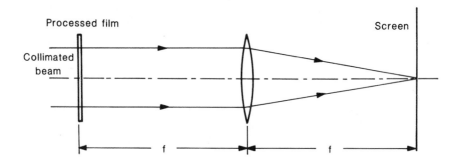

Fig. 14.3. Formation of Young's fringes by a film exposed to two laterally displaced speckle patterns.

where $h(u,v) \leftrightarrow I(x,y)$ and $k = 2\pi/\lambda$. The first term on the right hand side of Eq. (14.15) is merely the directly transmitted beam, which is brought to a point focus on axis and is of no interest. The intensity due to the second term is

$$\beta^2 T^2 |h(u,v)|^2 \, (1 + \cos k v y_0) \qquad (14.16)$$

which corresponds to the power spectrum of the speckle pattern modulated by a set of equally spaced interference fringes. These fringes are commonly referred to as Young's fringes, because they correspond to the fringes that would be obtained if any pair of corresponding points on the two speckle patterns were replaced by two coherent point sources.

14.7. Speckle photography

Speckle can be used to measure local displacements of a surface merely by illuminating it with a laser and taking two photographs of the surface on a fine-grain film or plate, one before and one after the movement [Archbold and Ennos, 1972].

14.7.1 In-plane displacement

If the object experiences a local in-plane displacement L the speckle pattern in the film plane undergoes a corresponding local displacement ML, where $M(<1)$ is the lateral magnification of the imaging lens. These local displacements can be measured by means of the simple spatial-filtering system shown in Fig. 14.4. In this arrangement, when the transparency is illuminated with collimated light, a system of Young's fringes is produced, as described in Section 14.5, in the back focal plane of the lens L_1. These fringes run at right angles to the displacement L, and their separation b satisfies the condition

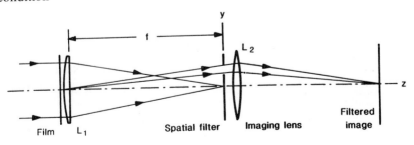

Fig. 14.4. Optical arrangement for spatial filtering of speckle photographs.

$$M|L|b/f = \lambda \qquad (14.17)$$

where f is the focal length of the lens. If, then, a small aperture is placed in the back focal plane of L_1 at a distance ρ from the optical axis along a line making an azimuthal angle θ, only those points whose components of displacement along this line satisfy the condition

$$M|L|\rho/f = m\lambda \qquad (14.18)$$

where m is an integer, will be reimaged by the lens L_2. Accordingly, the image is covered with fringes which are contours of equal displacement, the contour interval corresponding to a change in the displacement of the surface of $\lambda f/M\rho$.

14.7.2. Tilt

If the angle of incidence of the laser beam on the surface is i, and the angle at which it is viewed is r, then it can be shown that a tilt of the surface by an angle $\Delta\theta$ about an axis lying within the surface causes the speckle pattern also to move in the same direction through an angle.

$$\Delta\psi = (1 + \cos i/\cos r)\Delta\theta \qquad (14.19)$$

Local tilts of the surface can be measured by double-exposure photography with the camera lens moved away from focus by a known amount Δf [Archbold and Ennos, 1972; Tiziani, 1972]. The corresponding displacement of the speckles for near-normal illumination and viewing is then

$$a = \Delta\theta/2M\Delta f \qquad (14.20)$$

and can be measured as described earlier. A particular case of interest is when the camera is focused on a plane which is the image of the source 'reflected' in the surface; it is then possible to record only the tilts, independent of any lateral displacement [Gregory, 1976 a,b].

14.7.3. Vibration measurements

Time-averaged speckle photography can also be used to study in-plane vibrations [Tiziani, 1971; Archbold and Ennos, 1972]. We consider a surface vibrating in its own plane along the y axis so that the displacement of a point on it, at time t, from its original position is

$$y = a \sin \omega t \qquad (14.21)$$

The speckles are now recorded as streaks of length $2Ma$. The position probability density function of a speckle is

$$W(y) = (1/\pi Ma)\,[1-(y/Ma)^2]^{-\frac{1}{2}} \qquad (14.22)$$

for $-a < y < a$, and zero outside this range. Since the speckles spend most of the time near the two ends of this range, fringes similar to those produced by an object displacement of $2a$ will be seen in the back focal plane of the lens L_1, when the film is viewed in a setup such as that shown in Fig. 14.4.

The diffracted amplitude is, in this case, proportional to the Fourier transform of the position probability density function which is

$$\Omega(v) = \int_{-a}^{a} (1/\pi Ma)\,[1 - (y/Ma)^2]^{-\frac{1}{2}} \exp\,(-ivy)\mathrm{d}y$$

$$= J_0(Mav) \qquad (14.23)$$

where $v = (2\pi/\lambda f)y'$, y' being the distance from the optical axis. Accordingly, the diffracted intensity is proportional to $J_0^2\ (Mav)$.

14.7.4. Two-aperture speckle photography

In this technique [Duffy, 1972], a mask with two small openings is placed in front of the lens used to image the object on the film (see Fig. 14.5). If the apertures have a diameter d and are separated by a distance D along the y axis, speckles with an average size $\lambda z_i/d$ are formed, which are modulated by fine interference fringes running parallel to the x axis, with a spacing $\lambda z_i/D$. When a double exposure is made with this system, the interference fringes

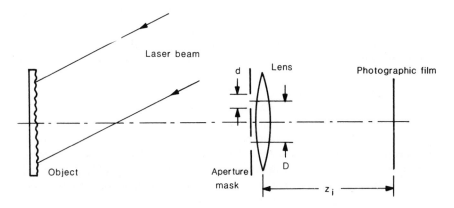

Fig. 14.5.   Optical arrangement for two-aperture speckle photography.

have maximum contrast in regions where the displacement of the surface of the object along the y axis is zero, or an integral multiple of $\lambda z_i/MD$. On the other hand, in regions where the surface displacement is $(m + \frac{1}{2})\,(\lambda z_i/MD)$, where m is an integer, the interference fringes recorded during the two exposures will be mutually displaced by half a fringe, so that the resultant contrast is zero.

If, then, the transparency is placed in a system such as that shown in Fig. 14.4, regions containing high-contrast fringes will diffract more light than those where the contrast of the fringes is less. Accordingly, with an aperture located at $y = \lambda f/D$, the image will exhibit fringes corresponding to displacements in the y direction.

$$L_y = m(\lambda z_i/MD) \tag{14.24}$$

Because of its simplicity, speckle photography is a very useful technique which complements holographic interferometry. The smallest displacements which can be measured correspond to a movement of the speckles in the film plane of a few times the speckle size, while the upper limit is set by overall decorrelation of the two speckle patterns. With an f/4 lens, for which the speckles are about 3 μm across, this would correspond to movements of the speckles of between 15 μm and 200 μm.

14.8. Speckle interferometry

Speckle interferometry differs from speckle photography in two main respects. The first is that it involves recording the speckle pattern formed by interference between the speckled image of the object and a uniform reference field or, more commonly, another speckle field. The second is that fringes are obtained due to local changes in the degree of correlation between two such speckle patterns. The sensitivity of the fringes to surface displacements is similar to that obtained with holographic interferometry.

14.9. Speckle interferometry with a uniform reference field

Consider a Michelson interferometer in which one of the mirrors has been replaced by a diffusely reflecting surface. The image of this surface is covered with speckles which are formed by interference of the speckle field at the surface and the reference plane wave. Any longitudinal movement Δz of the surface results in a change in the phase difference between the two fields

$$\Delta\phi = (2\pi/\lambda)2\Delta z \tag{14.25}$$

and a consequent change in the intensity distribution in the speckle pattern. With a slow movement of the surface, the speckle appears to twinkle.

Such a system can be used very effectively for studying vibrating objects [Archbold et al., 1969; Stetson, 1970b]. For this, an image of the object is formed by a lens with a large f-number so that the speckles are easily visible. When the object vibrates, the speckle in the vibrating areas where the pattern is continuously changing is effectively averaged and almost disappears, while the nodes stand out as regions of high-contrast speckle.

It is also possible to replace both the mirrors M_1, M_2 in a Michelson interferometer by scattering surfaces [Leendertz, 1970]. In this case, as discussed in Section 14.4, the coherent superposition of the speckle fields $a_1(x,y)$ and $a_2(x,y)$ in the images of the two surfaces yields a speckle field $a(x,y)$ whose intensity distribution $I(x,y)$ differs from the intensity distributions $I_1(x,y)$ and $I_2(x,y)$ in the component speckle fields but whose statistics are very similar to theirs. A movement Δz of one of the surfaces (say M_1) will result, as before, in the intensity distribution $I(x,y)$ changing. However, for a displacement $\Delta z = m\lambda/2$, where m is an integer, the resultant phase change is $\Delta\phi = 2m\pi$ and the pattern will once again be exactly the same as that when $\Delta z = 0$.

It is possible, therefore, to evaluate the displacement of M_1 from the degree of correlation of the speckle patterns produced in these two positions. One way of doing this is to record the first speckle pattern on a photographic film which is replaced, after processing, in its original position. If M_1 has not moved, the light transmitted through this film will be a minimum, since dark areas in the film will coincide with the bright speckles. However, if the surface M_1 undergoes a displacement which varies across the field, the amount of light transmitted will be a minimum only in those areas where the correlation between the two speckle patterns is complete. Dark fringes will then be seen across the image corresponding to surface displacements for which $\Delta\phi = 2m\pi$.

Another way of generating speckle-correlation fringes is to record both the speckle patterns on the same photographic film [Archbold et al., 1970]. With a high contrast material, the regions where the two speckle patterns differ will be very nearly opaque, because areas left unexposed by the dark speckles in one pattern will be filled in by the bright speckles in the other. However, regions where the speckle pattern has not changed will have a significant residual transmittance, due to the large number of points for which the illumination is zero for both exposures. With Polaroid film the fringes can be viewed almost immediately [Hariharan, 1978]. Higher contrast fringes can be obtained by moving the photographic film sideways, by a very small amount between the two exposures [Archbold et al., 1970; Butters and Leendertz, 1971a]. When the developed film is viewed in an optical processor, such as that shown in Fig. 14.3, Young's fringes are formed in the

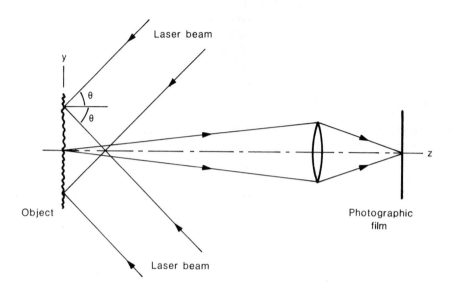

Fig. 14.6. Speckle interferometer for measuring in-plane displacements [Leendertz, 1970].

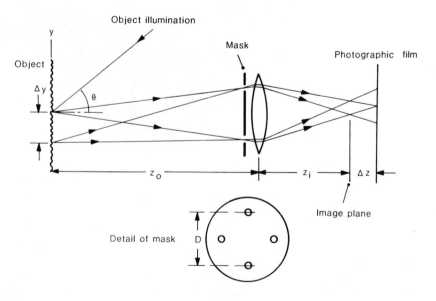

Fig. 14.7. Optical system for four-aperture speckle-shearing interferometry [Hung *et al.*, 1975].

back focal plane of L_1 only by those regions of the patterns which exhibit a strong correlation. Accordingly, if two apertures are positioned in the back focal plane of L_1, on the first maximum on either side of the optical axis, only those regions of the speckled images which exhibit a strong correlation will be seen in the final image.

14.10. In-plane displacements

Speckle interferometry is a particularly convenient way of measuring in-plane displacements [Leendertz, 1970]. For this, the surface is illuminated, as shown in Fig. 14.6 by two beams of coherent light incident at equal angles $\pm\theta$ to the normal to the surface. A speckle pattern is formed by coherent superposition of the two speckle fields produced by the two incident beams. The additional phase difference introduced between these component speckle fields at any point $P(x,y)$ by a displacement $L(x,y)$ of this point is

$$\Delta\phi = (k_1 - k_2) \cdot L(x,y) \qquad (14.26)$$

where k_1 and k_2 are the propagation vectors of the two incident waves. Equation (14.26) shows that this phase difference is proportional to the component of the displacement in the plane containing k_1 and k_2, normal to the bisector of the angle between them. Accordingly, the fringes obtained with a film on which two exposures have been made are contours of equal displacement along the y axis, maximum correlation of the speckle patterns occurring when

$$L_y = m\lambda/4\sin\theta \qquad (14.27)$$

where m is an integer.

14.11. Speckle-shearing interferometry

In speckle-shearing interferometry, a speckle pattern is produced by interference of two sheared images of the surface of the object. This permits direct measurements of the derivatives of the surface displacements.

Several optical arrangements have been used to produce the sheared images [Leendertz and Butters, 1973; Hung and Taylor, 1974; Hariharan, 1975d; Hung and Liang, 1979]. A particularly simple system described by Hung et al. [1975] which permits simultaneous measurements of the strain along two orthogonal directions is shown in Fig. 14.7. In this setup, a mask

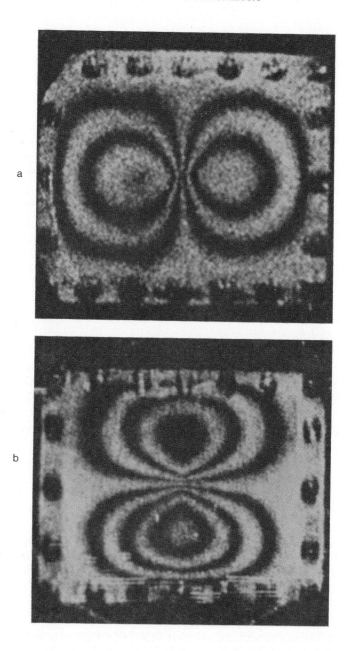

Fig. 14.8. Speckle-shearing interferograms of a centrally loaded rectangular plate showing (a) $(\partial L_z/\partial x)$, and (b) $(\partial L_z/\partial y)$. [Hung *et al.*, 1975].

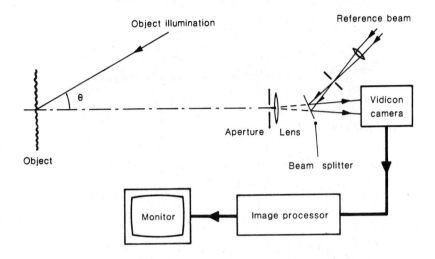

Fig. 14.9.   Basic system for electronic speckle-pattern interferometry.

containing four symmetrically placed apertures, two along the x axis and two along the y axis, is placed in front of the imaging lens. Two photographs of the surface illuminated with coherent light, before and after deformation, are recorded on a high resolution film which is located at a small distance Δz from the image plane. This produces four images of the object, one pair sheared along the x axis and the other sheared along the y axis by amounts equivalent to distances on the object

$$\Delta x = \Delta y = Dz_o \, \Delta z / z_i^2 \qquad (14.28)$$

where D is the separation of a pair of apertures and z_o and z_i are the distances of the object and the image, respectively, from the lens.

If the angle of illumination is small ($\theta \approx 0$), it can be shown that the additional phase difference between the first two speckle fields, which are sheared along the x axis, is

$$\Delta\phi(x) \approx (2\pi/\lambda)(\partial L_z / \partial x) \, (1 + \cos\theta)\Delta x \qquad (14.29)$$

while that between the other two speckle fields, which are sheared along the y axis is

$$\Delta\phi(y) \approx (2\pi/\lambda) \, (\partial L_z / \partial y) \, (1 + \cos\theta)\Delta y \qquad (14.30)$$

where L_z is the z component of the surface displacement of the object.

If the processed film is placed in a spatial filtering system, such as that shown in Fig. 14.4, the diffraction pattern seen in the back focal plane of L_1 will contain four bright spots, one on either side of the optical axis, and one just above and below it. With a spatial filter which permits selecting either the first or the second pair of spots, fringes will be obtained corresponding, respectively, to $\Delta\phi(x) = 2\pi m$ and $\Delta\phi(y) = 2\pi m$, where m is an integer. Speckle-shearing interferograms of a centrally loaded rectangular plate obtained with this technique, displaying contours of $(\partial L_z/\partial x)$ and $(\partial L_z/\partial y)$ are shown in Fig. 14.8.

14.12. Electronic speckle-pattern interferometry (ESPI)

The use of television cameras instead of photographic materials for speckle-pattern interferometry was first described by Butters and Leendertz [1971b] and, almost simultaneously, by Macovski et al. [1971]. A typical system used to study movements of an object along the line of sight is shown schematically in Fig. 14.9. The object is imaged by a lens, stopped down to about $f/16$, on a silicon target vidicon [Pedersen et al., 1974] on which is also incident a reference beam which diverges from a point located effectively at the centre of the lens aperture. The resulting image interferogram has a coarse speckle structure which can just be resolved by the television camera. The video signal from the television camera is electronically processed and then displayed on a television monitor. The theory of fringe formation and signal processing to obtain optimum fringe contrast has been studied by Slettemoen [1977, 1979], who has also described a modified system which uses a speckle reference beam [Slettemoen, 1980].

14.12.1. Displacement measurements

To measure displacements of the object, two speckled images recorded with the object in its initial and final positions can be stored and the difference extracted. Alternatively, a stored signal can be subtracted in real time from the signal from the camera. This can be done using a video tape recorder [Butters and Leendertz, 1971b], a scan converter memory tube [Løkberg et al., 1976] or a digital store [Nakadate et al., 1980a]. In either case, the resulting difference signal is enhanced by high-pass filtering and rectification before being fed to the television monitor. Regions in which the speckle pattern has not changed then give a zero difference signal and appear dark in the image, while regions where the speckle pattern has changed are covered with bright speckles. The dark fringes seen in the image are, therefore, contours of equal displacement defined by the condition

$$K \cdot L(x,y) = 2m\pi \qquad (14.31)$$

where m is an integer.

ESPI can also be used to measure in-plane deformation by replacing the imaging lens and the film in an optical arrangement such as that shown in Fig. 14.8, by a television camera [Denby and Leendertz, 1974]. It can also be used for speckle-shearing interferometry [Nakadate et al., 1980b].

14.12.2. Vibration analysis

One of the most valuable applications of ESPI has been to provide a real-time display of the vibration amplitude of an object. With a setup such as that shown in Fig. 14.9, the target of the television camera integrates the incident intensity over the period of a scan, which is about 40 ms. If the period of the vibration is much shorter than this, the variation of the contrast of the speckles with the vibration amplitude is given by the expression [Ek and Molin, 1971]

$$C = \{1 + 2\alpha J_0^2[K \cdot L(x,y)]\}^{1/2}/(1+\alpha) \qquad (14.32)$$

where α is the ratio of the intensities of the reference beam and the object beam, K is the sensitivity vector and $L(x,y)$ is the vibration amplitude at a point $P(x,y)$ on the object.

Since the reference beam contributes a uniform background which tends to mask the variations in the contrast of the speckle pattern, the signal from the camera is processed first to remove this dc component, after which it is low-pass filtered to remove electronic noise, rectified and displayed on the television monitor. Areas of the image corresponding to the zeros of the Bessel function, where the contrast of the speckle pattern is a minimum, appear as dark fringes on the monitor.

Because of the limited dynamic range available with ESPI, only about 10 fringes can be seen in this time-averaged mode of operation. The measuring range can be increased by phase modulation of the reference beam. This technique can also be used to obtain increased sensitivity and to determine the phase of the vibration [Løkberg and Høgmoen, 1976a,b; Høgmoen and Pedersen, 1977; Løkberg, 1979]. Another interesting possibility is to use stroboscopic illumination; in this case $\cos^2$ fringes are obtained, showing the deformation of the object between the pulses. Stroboscopic illumination even permits studying the vibrations of unstable subjects such as the human ear drum *in vivo* [Løkberg et al., 1979]. Transient events and vibrating objects can also be studied with a double-pulsed ruby laser [Cookson et al., 1978].

14.12.3. Contouring

Surface contouring can be carried out with ESPI [Jones and Butters, 1975] by recording two speckle images with wavelengths λ_1 and λ_2. In this case, the two speckle images exhibit a correlation peak when the optical path difference between the two beams satisfies the condition

$$p = m_1 \lambda_1 = m_2 \lambda_2 \tag{14.33}$$

where m_1 and m_2 are integers. Fringes are seen, therefore, at intervals given by the relation

$$\Delta p = \lambda_1 \lambda_2 / (\lambda_1 - \lambda_2) \tag{14.34}$$

This technique has obvious similarities to holographic contouring.

15
Stellar interferometry

A problem which has attracted the attention of astronomers for centuries is the measurement of stellar diameters. The resolution of conventional telescopes is inadequate for this purpose, since even the largest stars have angular diameters of the order of 10^{-2} second of arc. On the other hand, interferometric methods are very suitable for such measurements, since a star can be modelled as an incoherent circular source over which the intensity distribution follows some simple law. Some of these methods are described in this chapter.

15.1. Michelson's stellar interferometer

Since the dimensions of a star are very small compared with its distance from the earth, it follows from the van Cittert–Zernike theorem (see Section 3.5) that the complex degree of coherence between the light vibrations from the star reaching two points on the earth's surface is given by the normalized Fourier transform of the intensity distribution over the stellar disc. The angular diameter of the star can, therefore, be obtained from a series of measurements of the complex degree of coherence at points on the earth separated by different distances. This can be done, as discussed in Section 3.4, by observations of the visibility of the fringes in an interferometer which samples the wave field from the star at these points.

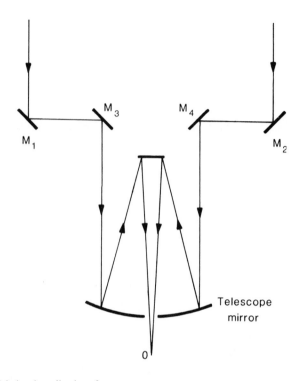

Fig. 15.1. Michelson's stellar interferometer.

In Michelson's stellar interferometer [Michelson and Pease, 1921], which is shown schematically in Fig. 15.1, four mirrors M_1, M_2, M_3, M_4 were mounted on a 6 m long support on the 2.5 m (100 inch) telescope at Mt Wilson. The light from the star received by the two mirrors M_1, M_2, whose spacing could be varied, was reflected by the fixed mirrors M_3, M_4, to the main telescope mirror, which brought the two beams to a focus at O.

In this arrangement, when the two images of the star are superimposed, and the two optical paths are equalized, straight, parallel interference fringes are seen, whose spacing is

$$\Delta x = \lambda f / d \tag{15.1}$$

where f is the focal length of the telescope and d is the separation of M_3 and M_4, while their visibility is

$$\mathscr{V} = |\mu_{12}| \tag{15.2}$$

where μ_{12} is the complex degree of coherence of the wave fields at M_1 and M_2. When M_1, M_2 are close together, $\mathscr{V}$ is nearly unity, but as D, the separation of M_1 and M_2, is increased, $\mathcal{V}$ decreases until, eventually, the fringes vanish.

If we assume the stellar disc to be a uniform circular source with an angular diameter 2α, the visibility of the fringes is, from Eq. (3.59),

$$\mathcal{V} = 2J_1(u)/u \qquad (15.3)$$

where $u = 2\pi\alpha D/\lambda$. The fringe visibility then varies with the separation of the mirrors as shown by the solid line in Fig. 15.2, dropping to zero when

$$D = 1.22 \, \lambda/2\alpha \qquad (15.4)$$

Measurements with Michelson's stellar interferometer present serious difficulties because of two very stringent requirements that must be met to ensure that the observed disappearance of the fringes is actually due to the finite diameter of the star. One is that the optical path difference between the two beams must be small compared to the coherence length of the light. The other is that the optical path difference must be stable to a fraction of a wavelength. The latter condition is very difficult to satisfy, even with a rigid structure, since rapid random changes are introduced in the two optical paths by atmospheric turbulence.

Because of these problems, an instrument with a baseline of 15 m built later on did not give consistent results and was finally abandoned. However, modern detection, control and data-handling techniques offer the possibility of overcoming these difficulties [Tango and Twiss, 1980]. At least two instruments of this type, designed to make measurements over baselines ranging from 100 m up to 1 km, are under construction [Liewer, 1979; Davis, 1979, 1984].

The optical system of one of these instruments, [Davis, 1984], is shown schematically in Fig. 15.3. Two coelostats C_1, C_2, on concrete plinths at the ends of the North–South baseline, direct light from the star via two beam-reducing telescopes (BRT) and an optical path-length compensator (OPLC) to the beam-splitter B. Error signals from two quadrant detectors (Q_g and Q) in each channel, viewing the image of the star, are used to control the coelostats and the two piezoelectric-actuated tilting mirrors (T), respectively, to keep these images exactly on the axis. Interference therefore takes place between two pairs of nominally parallel wavefronts leaving the beam-splitter B.

Two photon counting detectors D_1 and D_2 measure the total flux in a narrow spectral band in the two interference patterns within a sampling time of 1 – 10 ms. The signals from these two detectors will then be proportional to $(1 + |\mu_{12}| \cos \phi)$ and $(1 - |\mu_{12}| \cos \phi)$, where ϕ is a phase angle which varies randomly with time because of changes in the optical path through the atmosphere. However, the mirrors S which are mounted on piezoelectric

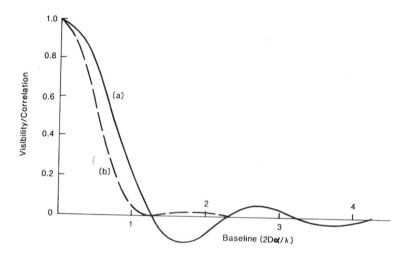

Fig. 15.2. Variation with the length of the baseline, for a uniform circular source, of (a) the visibility of the fringes in Michelson's stellar interferometer, and (b) the correlation in an intensity interferometer.

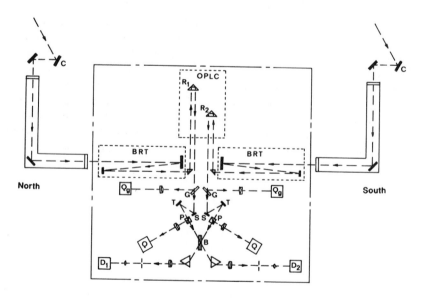

Fig. 15.3. Schematic diagram of a long-baseline stellar interferometer (Courtesy J. Davis, University of Sydney).

translators enable an additional phase difference of 90° to be introduced between the two interfering beams in alternate sampling periods. As a result, the signals from the two detectors during the next period will be proportional to $(1 + |\mu_{12}| \sin \phi)$ and $(1 - |\mu_{12}| \sin \phi)$. The average of the square of the difference between the two signals then gives $2|\mu_{12}|^2 < \cos^2 \phi + \sin^2 \phi >$. It is anticipated that this instrument will permit measurements of $|\mu_{12}|$ with an accuracy of 2%. It should therefore be possible to obtain values for the angular diameter of a star, using Eq. (15.3), from measurements at smaller mirror separations than that corresponding to the first zero of the visibility curve.

15.2. The intensity interferometer

The problems of maintaining equality of the two optical paths to less than a few wavelengths and minimizing the effects of atmospheric turbulence are eliminated in the intensity interferometer [Hanbury Brown and Twiss, 1956, 1957a,b, 1958a,b]. In this instrument, as shown schematically in Fig. 15.4, light from a star is focused on two photoelectric detectors, which are separated by a distance which can be varied, and the correlation between the fluctuations in the output currents from the two detectors is measured.

The fluctuations in the output currents from the two detectors consist of two components. One is the shot noise associated with the current, while the other, which is smaller and may be called the wave noise, is due to fluctuations in the intensity of the incident light. The shot noise in the two detectors is not correlated, but the wave noise exhibits a correlation which depends on the degree of coherence of the light at the two detectors.

To evaluate this correlation, we consider, as shown in Fig. 3.2, two points P_1, P_2, illuminated by an incoherent source. The size of the source and the separation of P_1 and P_2 are assumed to be very small compared to the distance of the source from P_1 and P_2. The intensities at P_1 and P_2 at time t can then be written as

$$I_1(t) = V_1(t) \, V_1^*(t) \qquad (15.5)$$

and

$$I_2(t) = V_2(t) \, V_2^*(t) \qquad (15.6)$$

where $V_1(t)$ and $V_2(t)$ are the analytic signals corresponding to the wave fields at P_1 and P_2. If the interferometer introduces a delay τ between these signals, the correlation between them is then

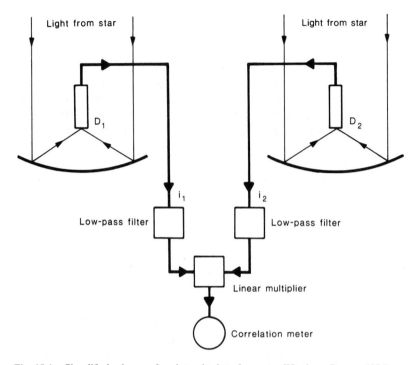

Fig. 15.4. Simplified scheme of an intensity interferometer [Hanbury Brown, 1974].

$$<I_1(t)I_2(t+\tau)> \; = \; <V_1(t)V_1{}^*(t)V_2(t+\tau)V_2{}^*(t+\tau)> \qquad (15.7)$$

However, since $V_1(t)$ and $V_2(t)$ are complex Gaussian processes, it follows from a theorem of Reed [1962] that

$$
\begin{aligned}
<V_1(t)&V_1{}^*(t)V_2(t+\tau)V_2{}^*(t+\tau)> \\
&= \; <V_1(t)V_1{}^*(t)><V_2(t+\tau)V_1{}^*(t+\tau)> \\
&\quad + <V_1(t)V_2{}^*(t+\tau)><V_1{}^*(t)V_2(t+\tau)> \qquad (15.8)
\end{aligned}
$$

Accordingly, from Eqs (3.7) and (3.19),

$$<I_1(t)I_2(t+\tau)> \; = \; I_1I_2 \; + \; |\Gamma_{12}(\tau)|^2 \qquad (15.9)$$

where I_1 and I_2 are the time-averaged intensities at P_1 and P_2, and $\Gamma_{12}(\tau)$ is the mutual coherence function of the wavefields at P_1 and P_2.

The second term on the right hand side of Eq (15.9) corresponds to the cross-correlation of $\Delta I_1(t)$ and $\Delta I_2(t)$, the fluctuations of intensity about the mean values I_1 and I_2, so that, from Eqs (15.9) and (3.20), we have

$$R_{12}(\tau) = \,<\Delta I_1(t)\Delta I_2(t+\tau)>$$
$$= |\Gamma_{12}(\tau)|^2$$
$$= I_1 I_2 |\gamma_{12}(\tau)|^2 \tag{15.10}$$

where $\gamma_{12}(\tau)$ is the complex degree of coherence. Equation (15.10) shows that the correlation observed at any given separation of the two detectors is proportional to the square of the modulus of the complex degree of coherence of the wave fields at these points. When $\tau = 0$, the normalized value of the correlation reduces to $|\mu_{12}|^2$; its variation with the length of the baseline for a uniform circular source is then shown by the broken line in Fig. 15.2.

It should be noted that the analysis leading to Eq. (15.10) applies only to linearly polarized light. When dealing with unpolarized light, it is necessary to take into account the fact that the orthogonal components of the field are uncorrelated. Equation (15.10) then becomes [Mandel, 1963]

$$R_{12}(\tau) = (1/2)I_1 I_2 |\gamma_{12}(\tau)|^2 \tag{15.11}$$

As we have seen in Section 15.1, the principal disadvantage of Michelson's stellar interferometer is that small, random changes in the optical paths traversed by the two beams, due to atmospheric turbulence, result in movements of the fringes which make observations difficult. These movements (see Section 3.4) correspond to changes in the phase of the complex degree of coherence. However, the intensity interferometer measures the square of the modulus of the complex degree of coherence, which is completely unaffected by changes in its phase. As a result, atmospheric turbulence has a negligible effect on the measured correlation in the intensity interferometer.

In addition, it can be shown that it is only necessary to maintain equality of the two paths to a few centimetres to obtain satisfactory results. Thus, if we define the Fourier transforms of the fluctuations of the electrical signals as

$$g_1(f) \leftrightarrow \Delta i_1(t) \tag{15.12}$$

and

$$g_2(f) \leftrightarrow \Delta i_2(t) \tag{15.13}$$

it follows from the Wiener–Khinchin theorem that

$$<\Delta i_1(t)\Delta i_2(t+\tau)> = \int_0^\infty g_1(f)\,g_2(f)\exp(-i2\pi f\tau)\mathrm{d}f$$
$$= \int_0^\infty S_{12}(f)\exp(-i2\pi f\tau)\mathrm{d}f \tag{15.14}$$

where $S_{12}(f)$ is the cross-spectral density of the fluctuations. Accordingly,

$$R_{12}(\tau)/R_{12}(0) = \int_0^\infty S_{12}(f) \exp(-i2\pi f\tau)df \Big/ \int_0^\infty S_{12}(f)df \quad (15.15)$$

so that the variation of the correlation with the delay is given by the normalized Fourier transform of the cross-spectral density of the fluctuations.

In this instrument, the band width of the electrical signals $\Delta i_1(t)$ and $\Delta i_2(t)$ at the correlator is such smaller than the optical bandwidth, and is determined by the low-pass filters in the two channels. If, therefore, we consider the simplest case, where the filters are identical and have a rectangular pass-band extending from $f = 0$ to $f = f_m$, Eq. (15.15) reduces to

$$R_{12}(\tau)/R_{12}(0) = \text{sinc}(f_m\tau) \quad (15.16)$$

Equation (15.16) shows that with a bandwidth of 100 MHz, the correlation drops to 90% of the maximum value for a delay corresponding to an optical path difference of 300 mm. It follows that the intensity interferometer is about 10^5 times less sensitive to the effects of optical path differences and time delays in the two paths than the Michelson stellar interferometer. As a result, the light collectors need not be finished to normal optical tolerances. In addition, it is possible to work with much longer baselines.

In a large instrument of this type [Hanbury Brown, 1964, 1974], the light collectors had a diameter of about 6.5 m and were made up of 252 hexagonal glass mirrors. These collectors were mounted on trucks running on a circular railway track with a diameter of 188 m. To follow a star, the two trucks moved around the track, while maintaining the desired separation. The beam from each collector after passing through an interference filter with a pass-band 8 nm wide centred on a wavelength of 433 nm, was focused on a photomultiplier.

The main drawback of this instrument is the very small signal available, because of the narrow bandwidth of the low-pass filters. This has limited its use to stars brighter than + 2.5 magnitude. Even with these, the correlation signal has to be integrated over a period of several hours to obtain a satisfactory S/N ratio. Measurements have been made successfully on 32 stars with angular diameters down to 0.42×10^{-3} second of arc.

15.3. Stellar speckle interferometry

The resolution of a telescope should increase, in theory, with its diameter. However, because of turbulence in the earth's atmosphere, there is little improvement in resolution when the aperture of the telescope exceeds

100 mm. Thus, even though the diffraction limited resolution of a 5 m tele-scope should be about 0.02 second of arc, the star images typically have an angular diameter of about 1 second of arc. Stellar speckle interferometry [Labeyrie, 1970, 1976] is a technique which can give, within certain lim-itations, diffraction-limited imaging of stellar objects, in spite of image de-gradation by the atmosphere and telescope aberrations.

Stellar speckle interferometry makes use of the fact that the image of a star in a large telescope, when observed under high magnification, exhibits a speckle structure [Texereau, 1963]. Individual speckles in such a pattern are actually equivalent to diffraction-limited images of the star [see Hariharan, 1972]. However, since the speckles are in continuous motion, short exposures ($\approx$ 20 ms) are essential to record these patterns. In addition, it is necessary to use a filter with a bandwidth of about 20 nm and to compensate for the dispersion of the atmosphere to obtain sharp images of the speckles.

The technique used by Labeyrie [1970] to extract a high-resolution image from a number of speckled images is shown schematically in Fig. 15.5. If $h(u,v)$ is the amplitude transmittance function of the perturbed atmosphere at a given instant, the corresponding distribution of the complex amplitude in the image of a point source is

$$H(x,y) \leftrightarrow h(u,v) \tag{15.17}$$

If $O(x,y)$ is the intensity distribution across a star, the intensity distribution $I(x,y)$ in a single, short-exposure speckle image can be written as

$$I(x,y) = O(x,y) * |H(x,y)|^2 \tag{15.18}$$

This image is recorded on a film which is processed so that the resultant amplitude transmittance $t_1 (x,y)$ is proportional to $I(x,y)$.

In the second stage, the transparency $t_1(x,y)$ is placed in the front focal plane of a lens L_2 and illuminated by a collimated beam of monochromatic light. The complex amplitude in the back focal plane of L_2 is then

$$\begin{aligned} a_2 (u,v) &= \mathscr{F}\{t_1(x,y)\} \\ &= \mathscr{F}\{O(x,y) * |H(x,y)|^2\} \\ &= \mathscr{F}\{O(x,y)\}R_{hh}(u,v) \end{aligned} \tag{15.19}$$

where $R_{hh} (u,v)$ is the autocorrelation function of $h(u,v)$. The intensity in the back focal plane of L_2 is, therefore,

$$I_2(u,v) = |\mathscr{F}\{O(x,y)\}|^2 |R_{hh}(u,v)|^2 \tag{15.20}$$

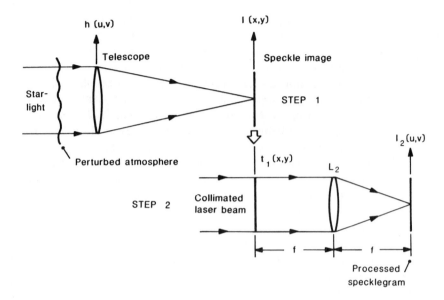

Fig. 15.5. Steps involved in processing a series of speckle images to obtain a diffraction-limited image of a star.

To improve the S/N ratio of the final result, many of these transformed specklegrams are averaged, for example, by making multiple exposures on the same photographic film. The resultant amplitude transmittance of this film is then

$$t_2(u,v) = <I_2(u,v)>$$
$$= | \mathscr{F}\{ O(x,y)\}|^2 <|R_{hh}(u,v)|^2 > \qquad (15.21)$$

The term within the pointed brackets on the right-hand side of Eq. (15.21) is the time-average of the transfer function of the system (atmosphere + telescope) for incoherent light, which can be obtained from measurements carried out under the same conditions on a neighbouring unresolved star. Since this transfer function has a finite positive value up to the diffraction limit of the telescope, the left-hand side of Eq. (15.21) can be divided by it to obtain $| \mathscr{F}\{O(x,y)\}|^2$. Fourier transformation of this result then gives the autocorrelation of the object,

$$| \mathscr{F}\{O(x,y)\}|^2 \leftrightarrow O(x,y) \star O(x,y) \qquad (15.22)$$

It should be noted that the angular resolution that can be obtained with speckle interferometry is limited by the aperture of the telescope and is, therefore, much less than what is possible with the long baseline interferometers described in Sections 15.1 and 15.2, though Labeyrie [1978] has proposed a system in which the speckle images from several telescopes are combined coherently to overcome this restriction. However, speckle interferometry has been applied successfully to a number of problems, including measurements of limb-darkening and oblateness of supergiant stars, with the 5 m (200 inch) telescope at Mt Palomar [Gezari *et al.*, 1972; Bonneau and Labeyrie, 1973]. It is probably most useful in the study of close double stars [Labeyrie *et al.*, 1974]. In this case, the speckled image corresponds to the superposition of two identical speckle patterns with a relative shift equal to the separation of the two stars. As a result, the transformed specklegram consists, as shown in Fig. 15.6, of a system of Young's fringes, from which the separation and the relative positions of the two stars can be inferred immediately [Labeyrie, 1974]. The scope and speed of the method have been progressively extended by the use of photon counters to record very faint images and digital techniques of image processing which permit averaging a very large number of transformed specklegrams [Beddoes *et al.*, 1976; Blazit *et al.*, 1977; Arnold *et al.*, 1979].

15.3.1. Elimination of aberrations

An interesting feature of the technique of stellar speckle interferometry is that, in principle, it is possible to obtain diffraction-limited resolution with a telescope of poor optical quality [Dainty, 1973]. The effective pupil function of the system (atmosphere + telescope) can be written as

$$h(u,v) = h_0(u,v) \, g(u,v) \qquad (15.23)$$

where $h_0(u,v)$ is the pupil function of the telescope and $g(u,v)$ varies in a random manner with time. It can then be shown that if $g(u,v)$ has a correlation distance that is small compared with the aperture of the telescope, the average transfer function for intensity obtained with a number of values of $g(u,v)$ contains, in addition to a spike at the origin, a term which corresponds to the aberration-free transfer function.

A similar technique proposed to obtain diffraction limited images with the Space Telescope is the speckle-rotation method [Lohmann and Weigelt, 1979; Weigelt, 1982] in which the telescope is rotated about its axis between successive exposures to obtain a series of speckled images; these images can then be processed to eliminate the aberrations of the telescope.

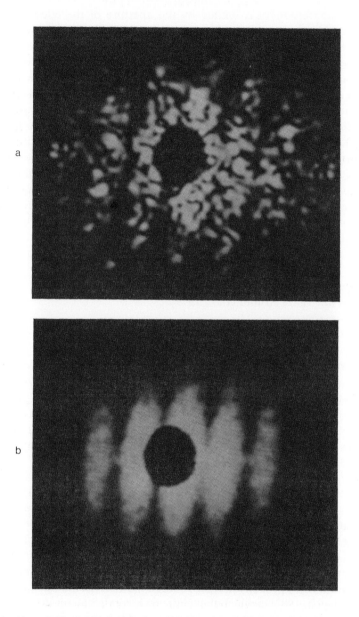

Fig. 15.6.   Young's fringes obtained from speckled images of a double star (ι Serpent) : (a) from a single exposure, and (b) the average of 300 exposures [Labeyrie,1974].

15.4. Speckle holography

A disadvantage of all the techniques of stellar interferometry discussed so far is that they do not give an image of the object, but only the autocorrelation of the image. This results in a 180° ambiguity in the position angles of the components in a group of stars. Accordingly, various methods have been developed for avoiding this ambiguity and reconstructing a diffraction limited image from a number of speckled images [Knox and Thompson, 1974; Ehn and Nisenson, 1975; Lynds *et al.*, 1976; Weigelt, 1977; von der Heide, 1978; Fienup, 1978; Bates and Cady, 1980]. Among these methods, one of the earliest, and one which is still of considerable interest, is speckle holography [Liu and Lohmann, 1973; Bates *et al.*, 1973; Gough and Bates, 1974].

The technique of speckle holography can be applied if there is, fairly close to $O(x,y)$, the astronomical object under study, an unresolved star $\delta(x-x_R, y-y_R)$. Speckle images are recorded of the total object

$$O_T(x,y) = O(x,y) + \delta(x-x_R, y-y_R) \qquad (15.24)$$

and these are processed using the Labeyrie technique to give its autocorrelation

$$
\begin{aligned}
O_T(x,y) \star O_T(x,y) = {} & \delta(x,y) + O(x,y) \star O(x,y) \\
& + O(-x+x_R, -y+y_R) \\
& + O(x-x_R, y-y_R)
\end{aligned}
\qquad (15.25)
$$

The distribution defined by Eq. (15.25) resembles a holographic image and consists of a central spot $\delta(x,y)$ on which is superposed the autocorrelation of the object, $O(x,y) \star O(x,y)$, and, away from the axis, on the opposite sides, the desired high-resolution image $O(x-x_R, y-y_R)$ and a conjugate image $O(-x+x_R, -y+y_R)$.

The application of this technique to a triple star, ADS 3358, is shown in Fig. 15.7, [Weigelt, 1978]. Figure 15.7(a) is one of 240 speckle images of ADS 3358 A–B–C which were recorded with a 1.8 m telescope, while Fig. 15.7(b) shows the optically produced average power spectrum and Fig. 15.7(c) is the reconstructed image of ADS 3358 A–B–C.

It should be noted that for this technique to be applied, the reference star and the object under study must lie within the same isoplanatic patch (or, in other words, within an area of space-invariant imaging); this limits the useful field to a few seconds of arc. At the same time, the distance from the reference star to the object $O(x,y)$ must be greater than 1.5 times the extent of $O(x,y)$. However, these limitations can be circumvented in many cases by suitable processing techniques [Weigelt, 1975].

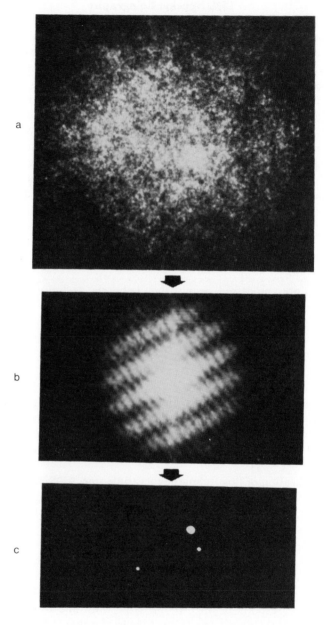

Fig. 15.7. Diffraction-limited imaging of the triple star ADS 3358 A–B–C by speckle holography: (a) speckle image; (b) averaged power spectrum, and (c) reconstructed image [Weigelt, 1978].

15.5. Speckle masking

An alternative method of eliminating the 180° ambiguity which does not require a reference point source is speckle masking [Weigelt and Wernitzer, 1983; Lohmann, Weigelt and Wernitzer, 1983]. This technique involves evaluating the third-order autocorrelation function of the image

$$<[I_n(x)\, I_n(x-m)] \star I_n(x)> \qquad (15.26)$$

where x and m are vectors in the image plane and $I_n(x)$ is the intensity distribution in the nth speckle image. The third-order autocorrelation function of the object is obtained from this by subtracting a set of bias terms. We have

$$\begin{aligned}
[O(x)\, O(x-m)] \star O(x) = \ & <[I_n(x)\, I_n(x-m)] \star I_n(x)> \\
& - <[I_m(x)\, I_n(x-m)] \star I_n(x)> \\
& - <[I_n(x)\, I_m(x-m)] \star I_n(x)> \\
& - <[I_n(x)\, I_n(x-m)] \star I_m(x)> \\
& + 2<[I_n(x)\, I_k(x-m)] \star I_m(x)> \qquad (15.27)
\end{aligned}$$

If, then, the masking vector m is chosen so that

$$O(x)\, O(x-m) \approx \delta(x) \qquad (15.28)$$

the left hand side of Eq. (15.27) reduces to

$$\begin{aligned}
[O(x)\, O(x-m)] \star O(x) &= \delta(x) \star O(x) \\
&= O(x) \qquad (15.29)
\end{aligned}$$

giving the intensity distribution across the object. Typically, in the case of measurements on a double star, the masking vector is equal to the vector joining the images of the two stars and can be evaluated from measurements of the autocorrelation using the Labeyrie technique.

15.6. Rotation-shearing interferometry

This technique [Mertz, 1965] has also been used for holography with incoherent illumination [Stroke and Restrick 1965; Lowenthal et al., 1969] and involves recording the interference pattern produced by two images of the telescope pupil which are rotated with respect to each other. This can be done by means of an interferometer which introduces a rotational shear [Armitage and Lohmann, 1965; Breckenridge, 1972; Roddier et al., 1978]. With a 180° rotational shear, each point of the object produces two images, located symmetrically with respect to the axis, which give rise to a system of straight parallel fringes running at right angles to the line joining these images, and with a spacing inversely proportional to their separation. With an incoherent object, the intensities due to all these points will add to produce a pattern which is the Fourier transform of the intensity distribution in the object. A recording of this pattern, when illuminated with a collimated beam, will reconstruct, in its far field, an image of the original object.

A problem in holography with incoherent illumination is that the S/N ratio is inversely proportional to the square root of the number of pixels in the image. However, this problem does not arise in astronomical imaging, where good results can be obtained as long as the number of resolved pixels is smaller than the number of speckles in the interferogram.

A major advantage of this technique is that aberrations of the telescope objective affect all the fringe systems in the same manner. As a result, it is possible to correct for these aberrations by recording a hologram of an unresolved star. After subtraction of the phase errors derived from this hologram, a good image is obtained. In the same manner, the effects of atmospheric turbulence can be eliminated by recording a number of interferograms with a short enough exposure time and averaging the fringe positions [Roddier, 1979].

Appendices

APPENDIX A1: Two-dimensional linear systems

The distribution of the complex amplitude in an image can be expressed by means of the two-dimensional Fourier transform as a function of two orthogonal spatial frequencies [Goodman, 1968]. As a result, the original light wave is decomposed into component plane waves whose direction cosines correspond to the various spatial frequencies in the image. The propagation of these components through an optical system can then be analysed using the concepts of communication theory.

A1.1. The Fourier transform

The two-dimensional Fourier transform of $g(x,y)$ is defined as

$$\mathscr{F}\{g(x,y)\} = \int_{-\infty}^{\infty} \int_{-\infty}^{\infty} g(x,y) \exp\left[-i2\pi(\xi x + \eta y)\right] \mathrm{d}x\mathrm{d}y$$

$$= G(\xi,\eta) \tag{A1.1}$$

Similarly, the inverse Fourier transform of $G(\xi,\eta)$ is defined as

$$\mathscr{F}^{-1}\{G(\xi,\eta)\} = \int_{-\infty}^{\infty} \int_{-\infty}^{\infty} G(\xi,\eta) \exp\left[i2\pi(\xi x + \eta y)\right] \mathrm{d}\xi\mathrm{d}\eta$$

$$= g(x,y) \tag{A1.2}$$

Equations (A1.1) and (A1.2) can be written symbolically as

$$g(x,y) \leftrightarrow G(\xi,\eta) \tag{A1.3}$$

Some important theorems are:

a. The linearity theorem

$$\mathcal{F}\{ag(x,y) + bh(x,y)\} = a\, G(\xi,\eta) + b\, H(\xi,\eta) \tag{A1.4}$$

where a and b are constants, and $g(x,y){\leftrightarrow}G(\xi,\eta)$, $h(x,y){\leftrightarrow}H(\xi,\eta)$

b. The shift theorem

$$\mathcal{F}\{g(x-a,y-b)\} = G(\xi,\eta)\, \exp[-i2\pi(\xi a + \eta b)] \tag{A1.5}$$

c. The similarity theorem

$$\mathcal{F}\{g(ax,by)\} = (1/|ab|)\, G(\xi/a,\eta/b) \tag{A1.6}$$

d. Rayleigh's theorem

$$\int_{-\infty}^{\infty}\int_{-\infty}^{\infty} |g(x,y)|^2\, dxdy = \int_{-\infty}^{\infty}\int_{-\infty}^{\infty} |G(\xi,\eta)|^2\, d\xi d\eta \tag{A1.7}$$

A1.2. Convolution and correlation

The output from any linear system is the convolution of the input and the instrument function. In two dimensions, the convolution of two functions, $g(x,y)$ and $h(x,y)$ is

$$f(x,y) = \int_{-\infty}^{\infty}\int_{-\infty}^{\infty} g(u,v)\, h(x-u,\, y-v)\, dudv \tag{A1.8}$$

which can also be written as

$$f(x,y) = g(x,y) * h(x,y) \tag{A1.9}$$

where the symbol $*$ denotes the convolution operation.

The cross-correlation of two functions, $g(x,y)$ and $h(x,y)$ is

$$c(x,y) = \int_{-\infty}^{\infty}\int_{-\infty}^{\infty} g^*(u,v)\, h(x+u,\, y+v)\, dudv) \tag{A1.10}$$

where $g^*(u,v)$ is the complex conjugate of $g(u,v)$. This can be written as

$$c(x,y) = g(x,y) \star h(x,y) \tag{A1.11}$$

where the symbol $\star$ denotes the correlation operation. A comparison of Eq. (A1.10) with Eq. (A1.8) shows that the cross-correlation can also be expressed as a convolution

$$c(x,y) = g^*(x,y) * h(-x,-y) \tag{A1.12}$$

The autocorrelation of a function $g(x,y)$ is then

$$a(x,y) = \int_{-\infty}^{\infty}\int_{-\infty}^{\infty} g^*(u,v)\, g(x+u,\, y+v)\, dudv$$

$$= g(x,y) \star g(x,y) \tag{A1.13}$$

Two useful results which follow are
a. The convolution theorem

If $g(x,y) \leftrightarrow G(\xi,\eta)$ and $h(x,y) \leftrightarrow H(\xi,\eta)$

$$\mathscr{F}\{g(x,y) * h(x,y)\} = G(\xi,\eta)\, H(\xi,\eta) \qquad (A1.14)$$

b. The autocorrelation (Wiener-Khinchin) theorem

$$\mathscr{F}\{g(x,y) \star g(x,y)\} = |G(\xi,\eta)|^2 \qquad (A1.15)$$

A1.3. The Dirac delta function

The two-dimensional Dirac delta function is a convenient representation of a point source. The
delta function takes the values

$$\delta(x,y) = \infty, (x = 0, \text{ and } y = 0)$$
$$\delta(x,y) = 0, (x \neq 0, \text{ or } y \neq 0) \qquad (A1.16)$$

and its integral is unity.

By definition, convolution of a function with the Dirac delta function yields the original
function, so that, in two dimensions,

$$\int_{-\infty}^{\infty} \int_{-\infty}^{\infty} f(u,v)\, \delta(x-u, y-v)\, du dv = f(x,y) \qquad (A1.17)$$

APPENDIX A2:   Random functions

The correlation function, as normally defined, cannot be used to study randomly varying quantities, since the correlation integral does not have a finite value. The cross-correlation of two stationary random functions $g(t)$ and $h(t)$ is therefore written [Papoulis, 1965] as

$$
\begin{aligned}
R_{gh}(\tau) &= \underset{T\to\infty}{\text{Lim}} \frac{1}{2T} \int_{-T}^{T} g^*(t)\, h(t+\tau)\, dt \\
&= \langle g^*(t)\, h(t+\tau)\rangle
\end{aligned}
\tag{A2.1}
$$

The autocorrelation function of a random function; say $g(t)$, is therefore

$$
R_{gg}(\tau) = \langle g^*(t)g(t+\tau)\rangle
\tag{A2.2}
$$

The power spectrum $S(\omega)$ of $g(t)$ is then defined as the Fourier transform of its autocorrelation function

$$
S(\omega)\leftrightarrow R_{gg}(\tau)
\tag{A2.3}
$$

APPENDIX A3:   The Fresnel–Kirchhoff integral

If, as shown in Fig. A3.1, a plane wave with an amplitude a is incident on an object with an amplitude transmittance $t(x_1,y_1)$ located in the plane $z = 0$, the complex amplitude at a point $P(x,y,z)$ is given by the Fresnel-Kirchhoff integral [Born and Wolf, 1980]

$$a(x,y,z) = (ia/\lambda) \int_{-\infty}^{\infty} \int_{-\infty}^{\infty} t(x_1,y_1)$$

$$\times \frac{\exp\{(-i2\pi/\lambda)[(x-x_1)^2 + (y-y_1)^2 + z^2]^{1/2}\}}{[(x-x_1)^2 + (y-y_1)^2 + z^2]^{1/2}}$$

$$\times \cos\theta \, dx_1 \, dy_1 \tag{A3.1}$$

When z is much greater than $(x-x_1)$ and $(y-y_1)$, $\cos\theta \approx 1$. We can then omit a factor $\exp(-i\,2\pi/\lambda z)$, which only affects the overall phase, and write Eq. (A3.1) as

$$a(x,y,z) = (ia/\lambda z) \int_{-\infty}^{\infty} \int_{-\infty}^{\infty} t(x_1,y_1)$$

$$\times \exp\{(-i\pi/\lambda z)[(x-x_1)^2 + (y-y_1)^2]\}dx_1 \, dy_1$$

$$= (ia/\lambda z) \int_{-\infty}^{\infty} \int_{-\infty}^{\infty} t(x_1,y_1) \, \exp[(-i\pi/\lambda z)(x^2+y^2)]\exp[(-i\pi/\lambda z)(x_1^2+y_1^2)]$$

$$\times \exp\{i2\pi[x_1(x/\lambda z) + y_1(y/\lambda z)]\} \, dx_1 \, dy_1 \tag{A3.2}$$

If, in addition, z is in the far field of the object, in which case,

$$z \gg (x_1^2 + y_1^2)/\lambda \tag{A3.3}$$

and if we set

$$\xi = x/\lambda z, \eta = y/\lambda z \tag{A3.4}$$

Eq. (A3.2) reduces to

$$a(x,y,z) = (ia/\lambda z) \exp[(-i\pi/\lambda z)(x^2 + y^2)]$$

$$\times \int_{-\infty}^{\infty} \int_{-\infty}^{\infty} t(x_1,y_1) \exp[i2\pi(\xi x_1 + \eta y_1)] \, dx_1 \, dy_1$$

$$= (ia/\lambda z) \exp[(-i\pi/\lambda z)(x^2 + y^2)] \, T(\xi,\eta) \tag{A3.5}$$

where

$$t(x_1,y_1) \leftrightarrow T(\xi,\eta) \tag{A3.6}$$

Accordingly, the complex amplitude in the plane of observation is given by the Fourier transform of the amplitude transmittance of the object, multiplied by a spherical phase factor. It can be shown that if the object transparency is placed in the front focal plane of a lens, and the plane of observation is located in its back focal plane, the exact Fourier transform is obtained. This follows because, in this case, z is effectively equal to infinity, and the phase factor $\exp[(-i\pi/\lambda z)(x^2+y^2)]$ reduces to unity.

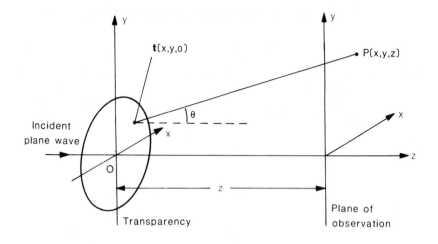

Fig. A3.1.　Derivation of the Fresnel–Kirchhoff integral.

APPENDIX A4: Reflection and transmission at a surface

A4.1. The Fresnel formulas

Consider a plane monochromatic wave incident, as shown in Fig. A4.1, on the interface between two isotropic transparent media with refractive indices n_1 and n_2. If θ_1 and θ_2 are the angles of incidence and refraction, the coefficients of reflection and transmission for amplitude are given by the Fresnel formulas, which can be written as

$$r_{\parallel} = \frac{\tan(\theta_1 - \theta_2)}{\tan(\theta_1 + \theta_2)} \tag{A4.1}$$

$$t_{\parallel} = \frac{2 \sin \theta_2 \cos \theta_1}{\sin(\theta_1 + \theta_2) \cos(\theta_1 - \theta_2)} \tag{A4.2}$$

for light polarized with its electric vector in the plane of incidence (the p – component), and

$$r_{\perp} = -\frac{\sin(\theta_1 - \theta_2)}{\sin(\theta_1 + \theta_2)} \tag{A4.3}$$

$$t_{\perp} = \frac{2 \sin \theta_2 \cos \theta_1}{\sin(\theta_1 + \theta_2)} \tag{A4.4}$$

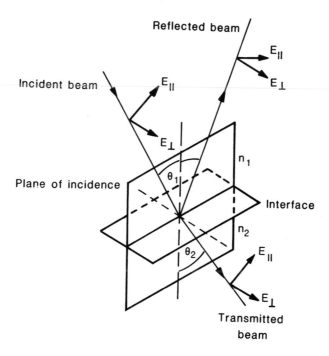

Fig. A4.1. Incident, transmitted and reflected fields at an interface ($n_1 > n_2$) when $\theta_1 + \theta_2 < \pi/2$.

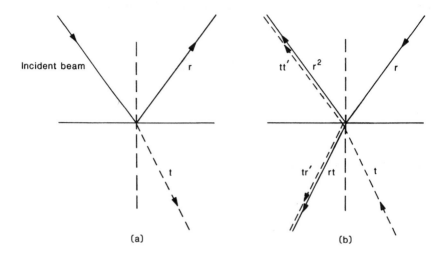

Fig. A4.2. Derivation of the Stokes relations.

for light polarized with the electric vector perpendicular to the plane of incidence (the s — component). At the Brewster angle, which is defined by the condition $r_s = 0$,

$$\theta_1 = \pi/2 - \theta_2 \tag{A4.5}$$

and

$$\tan \theta_1 = n_2/n_1 \tag{A4.6}$$

Since, below the critical angle, $t_\parallel$ and $t_\perp$ are always positive, there is no change in the phase of the transmitted wave. On the other hand, $r_\perp$, which is positive when $n_1 > n_2$, is negative when $n_1 < n_2$ indicating, in the latter case, a phase shift of π. In the case of $r_\parallel$ the phase shifts at small angles of incidence are the same as those for $r_\perp$ (the difference in their signs arises from the choice of the coordinate axes), but there is a change in sign at the Brewster angle.

A4.2. The Stokes relations

Let r and t be the reflected and transmitted amplitudes, respectively, for a wave of unit amplitude incident, as shown in Fig. A4.2 (a), on the interface between two transparent media. If, then, as shown in Fig. A4.2 (b), the direction of the reflected ray is reversed, it will give rise to a reflected component with an amplitude r^2 and a transmitted component with an amplitude rt. Similarly, if the direction of the transmitted ray is reversed, it will give rise to a reflected component with an amplitude tr' and a transmitted component with an amplitude tt', where r' and t' are, respectively, the reflectance and transmittance for amplitude for a ray incident on the interface from below. If there are no losses at the interface

$$r^2 + tt' = 1 \tag{A4.7}$$

and

$$rt + tr' = 0 \tag{A4.8}$$

Equations (A4.7) and (A4.8) lead to the Stokes relations

$$tt' = 1 - r^2 \tag{A4.9}$$

and

$$r' = -r \tag{A4.10}$$

It follows from Eq. (A4.10) that the phase shifts on reflection at the two sides of the interface differ by π.

APPENDIX A5:   The Jones calculus

Various methods of describing polarized beams are available, but problems involving coherent beams, in which their relative amplitudes and phases are important, can be handled most conveniently by means of the Jones calculus [Jones, 1941].

In the Jones calculus, the characteristics of a polarized beam are described by means of a two-element column vector known as the Jones vector. For a plane monochromatic light wave propagating along the z axis, which has components $E_x = |a_x| \cos(\omega t + \phi_x)$ and $E_y = |a_y| \cos(\omega t + \phi_y)$ along the x and y directions, respectively, the Jones vector takes the form

$$A = \begin{bmatrix} E_x \\ E_y \end{bmatrix} \tag{A5.1}$$

Since, with monochromatic light, the variation with time of E_x and E_y is not of interest, (A5.1) is often written in the form

$$A = \begin{bmatrix} a_x \\ a_y \end{bmatrix} = \begin{bmatrix} |a_x| \exp(i\phi_x) \\ |a_y| \exp(i\phi_y) \end{bmatrix} \tag{A5.2}$$

This is known as the full Jones vector.
The intensity at a point is then

$$\begin{aligned} I &= a_x a_x{}^* + a_y a_y{}^* \\ &= A^\dagger A \end{aligned} \tag{A5.3}$$

The intensity is, therefore, the product of the Jones vector and its Hermitian conjugate.

For convenience in calculations, the normalized form of the Jones vector is commonly used. This is obtained by multiplying the full Jones vector by a complex scalar that reduces the intensity to unity.

For example, for light that is linearly polarized at 45°, the amplitudes of the x and y components are the same, and the full Jones vector can be written as

$$\begin{bmatrix} a \\ a \end{bmatrix} \tag{A5.4}$$

The corresponding normalized vector is

$$2^{-1/2} \begin{bmatrix} 1 \\ 1 \end{bmatrix} \tag{A5.5}$$

Optical elements which modify the state of polarization of the beam are described by matrices containing four elements, known as Jones matrices. The Jones vector of the incident beam is multiplied by the Jones matrix of the optical element to obtain the Jones vector of the emerging beam.

Thus, for example, the Jones matrix for a half-wave plate with its axis vertical or horizontal is

$$\begin{bmatrix} 1 & 0 \\ 0 & -1 \end{bmatrix} \tag{A5.6}$$

If a beam linearly polarized at 45° is incident on it, the Jones vector for the emerging beam is

$$\begin{bmatrix} 1 & 0 \\ 0 & -1 \end{bmatrix} 2^{-\frac{1}{2}} \begin{bmatrix} 1 \\ 1 \end{bmatrix} = 2^{-\frac{1}{2}} \begin{bmatrix} 1 \\ -1 \end{bmatrix} \tag{A5.7}$$

which is a beam linearly polarized at − 45°.

An important element in interferometry is a metal film (a mirror or a beamsplitter). Oblique reflection at such a film introduces a phase shift between the components of the incident beam polarized parallel and perpendicular to the plane of incidence. As a result, the amplitude reflection coefficients $r_{\parallel}$ and $r_{\perp}$ are complex, and the film can be represented by the Jones matrix

$$M = \begin{bmatrix} r_{\parallel} & 0 \\ 0 & r_{\perp} \end{bmatrix} \tag{A5.8}$$

Account must be taken, of course, of the change from right-hand to left-hand coordinates on reflection.

In an optical system in which the beam passes through several elements in succession, the Jones vector of the beam leaving the system can be obtained by writing down the Jones matrices of the various elements and carrying out the multiplication in the proper sequence. Alternatively, where the state of polarization of the incident beam is to be varied, it is possible to compute the overall matrix for the system and then multiply the Jones vectors for the various states of the incident beam by this matrix.

The use of the Jones calculus has been described in detail by Shurcliff [1962], Shurcliff and Ballard [1964], and Jerrard [1982], who have also listed the Jones vectors for various forms of polarization, and the Jones matrices of the most important types of polarizers and retarders.

References

Abelès, F. (1950). *J. Phys. Radium.* **11**, 307-309.

Abelès, F. (1967). *In* "Advanced Optical Techniques" (A.C.S. van Heel, ed.), pp. 143-188. North-Holland, Amsterdam.

Abramson, N. (1969a). *Optik* **30**, 56-71.

Abramson, N. (1969b). *Appl. Opt.* **8**, 1235-1240.

Abramson, N. (1970a). *Appl. Opt.* **9**, 97-101.

Abramson, N. (1970b). *Appl. Opt.* **9**, 2311-2320.

Abramson, N. (1971). *Appl. Opt.* **10**, 2155-2161.

Abramson, N. (1972). *Appl. Opt.* **11**, 1143-1147.

Abramson, N. (1974). *Appl. Opt.* **13**, 2019-2025.

Abramson, N. (1975). *Appl. Opt.* **14**, 981-984.

Abramson, N. (1977). *Appl. Opt.* **16**, 2521-2531.

Abramson, N. and Bjelkhagen, H. (1973). *Appl. Opt.* **11**, 2792-2796.

Abramson, N. and Bjelkhagen, H. (1979). *Appl. Opt.* **18**, 2870-2880.

Aiki, K., Nakamura, M., Kuroda, T., Umeda, J., Ito, R., Chinone, N., and Maeda, M. (1978). *IEEE J. Quantum Electron.* **QE-14**, 89-94.

Aleksandrov, E.G. and Bonch-Bruevich, A.M. (1967). *Sov. Phys. Tech. Phys.* **12**, 258-265.

Aleksoff, C.C. (1971). *Appl. Opt.* **10**, 1329-1341.

Ammann, E.O. and Chang, I.C. (1965). *J. Opt. Soc. Am.* **55**, 835-841.

Archbold, E., Burch, J.M. and Ennos, A.E. (1970). *Opt. Acta* **17**, 883-898.

Archbold, E., Burch, J.M. Ennos, A.E. and Taylor, P.A. (1969). *Nature* **222**, 263-265.

Archbold, E., and Ennos, A.E. (1972) *Opt. Acta.* **19**, 253-271.

Armitage, J.D., Jr., and Lohmann, A. (1965). *Opt. Acta.* **12**, 185-192.

Armstrong, W.T., and Forman, P.R. (1977). *Appl. Opt.* **16**, 229-232.

Arnold, S.J., Boksenberg, A., and Sargent, W.L.W. (1979). *Astrophys. J.* **234**, L159-L163.

Ashby, D.E.T.F., and Jephcott, D.F. (1963). *Appl. Phys. Lett.* **3**, 13-16.

Baird, K.M. (1954). *J. Opt. Soc. Am.* **44**, 11-13.

Baird, K.M. (1963). *Appl. Opt.* **2**, 471-479.

Baird, K.M. (1981). *J. Phys. Colloq. (France).* **42 (C-8)**, 485-494.

Baird, K.M. (1983a). *Physics Today* **36** (1), 52-57.

Baird, K.M. (1983b). *In* "Quantum Metrology and Fundamental Physical Constants" (P.H. Cutler and A.A. Lucas, eds), pp.143-162. Plenum, New York.

Baird, K.M., and Hanes, G.R. (1974). *Rep. Progr. Phys.* **37**, 927-950.

Baird, K.M., and Howlett, L.E. (1963). *Appl. Optics.* **2**, 455-463.

Baker, D.J., Steed, A.J., and Stair, A.T.Jr., (1973). *J. Geophys. Res.* **78**, 8859-8863.

Balhorn, R., Kunzmann, H., and Lebowsky, F. (1972). *Appl. Opt.* **11**, 742-744.

Ballard, G.S. (1968). *J. Appl. Phys.* **39**, 4846-4848,

Banning, M. (1947). *J. Opt. Soc. Am.* **37**, 792-797.

Bar-Joseph, I., Hardy, A., Katzir, Y., and Silberberg, Y. (1981). *Opt. Lett.* **6**, 414-416.

Barrell, H., and Sears, J.E. (1939). *Phil. Trans. Roy. Soc.* **A238**, 1-64.

Bartelt, H.O., and Jahns, J. (1979). *Opt. Commun.* **30**, 268-274.

Bates, J.B. (1976). *Science* **191**, 31-37.

Bates, R.H.T., and Cady, F.M. (1980). *Opt. Commun.* **32**, 365-369.

Bates, R.H.T., Gough, P.T., and Napier, P.J. (1973). *Astron. and Astrophys.* **22**, 319-320.

Bates, W.J. (1947). *Proc. Phys. Soc.* **59**, 940-950.

Bay, Z., Luther, G.G., and White, J.A. (1972). *Phys. Rev. Lett.* **29**, 189-192.

Beddoes, D.R., Dainty, J.C., Morgan, B.L., and Scaddan, R.J. (1976). *J. Opt. Soc. Am.* **66**, 1247-1251.

Bell, R.J. (1972). "Introductory Fourier Transform Spectroscopy". Academic Press, New York.

Bennett, H.E., and Bennett, J.M. (1967). *In* "Physics of Thin Films" (G. Hass and R.E. Thun, eds), Vol. IV, pp.1-96. Academic Press, New York.

Bennett, S.J., Ward, R.E., and Wilson, D.C. (1973). *Appl. Opt.* **12**, 1406.

Beran, M.J., and Parrent, G.B., Jr. (1964). "Theory of Partial Coherence". Prentice-Hall, Englewood Cliffs.

Bergh, R.A., Lefrere, H.C., and Shaw, H.J. (1981). *Opt. Lett.* **6**, 198-200.

Berning, P.H., and Turner, A.F. (1957). *J. Opt. Soc. Am.* **47**, 230-239.

Biddles, B.J. (1969). *Opt. Acta.* **16**, 137-157.

Birch, K.G. (1973). *J. Phys. E: Sci. Instrum.* **6**, 1045-1048.

Blazit, A., Bonneau, D., Koechlin, L., and Labeyrie, A. (1977). *Astrophys. J.* **214**, L79-L84.

Bloom, A.L. (1968). "Gas Lasers". John Wiley, New York.

Bloom, A.L. (1974). *J. Opt. Soc. Am.* **64**, 447-452.

Bonneau, D., and Labeyrie, A. (1973). *Astrophys. J.* **181**, L1-L4.

Born M., and Wolf, E. (1980). "Principles of Optics". Pergamon Press, Oxford.

Bouchareine, P., and Connes, P. (1963). *J. de Phys.* **24**, 134-138.

Bourdet, G.L., and Orszag, A.G. (1979). *Appl. Opt.* **18**, 225-227.

Boyd, G.D. and Gordon, J.P. (1961). *Bell Syst. Tech. J.* **40**, 489-508.

Boyd, G.D. and Kogelnik, H. (1962). *Bell Syst. Tech. J.* **41**, 1347-1369.

Bradley, D.J., and Mitchell, C.J. (1968). *Phil. Trans. Roy. Soc.* **A263**, 209-223.

Bragg, W.L. (1939). *Nature* **143**, 678.

Breckenridge, J.B. (1972). *Appl. Opt.* **11**, 2996-2998.

Briers, J.D. (1972). *Opt. and Laser Tech.* **4**, 28-41.

Briers, J.D. (1976). *Opt. Quant. Electron.* **8**, 469-501.

Brillet, A., and Cérez, P. (1981). *J. Phys. Colloq. (France).* **42 (C-8)**, 73-82.

Brillet, A., and Hall, J.L. (1979). *Phys Rev. Lett.* **42**, 549-552.

Brooks, R.E., Heflinger, L.O., and Wuerker, R.F. (1965). *Appl. Phys. Lett.* **7**, 248-249.

Brossel, J. (1947). *Proc. Phys. Soc.* **59**, 224-234.

Brown, B.R. and Lohmann, A.W. (1966). *Appl. Opt.* **5**, 967-969.

Brown, B.R., and Lohmann, A.W. (1969). *IBM J. Res. Dev.* **13**, 160–167.
Brown, D.S. (1959). *In* "Interferometry: N.P.L. Symposium No. 11" pp. 253–256. Her Majesty's Stationery Office, London.
Brown, N. (1981). *Appl. Opt.* **20**, 3711–3714.
Brown, R. Hanbury (1964). *Sky. Telesc.* **28**, 64–69.
Brown, R. Hanbury (1974). "The Intensity Interferometer". Taylor and Francis, London.
Brown, R. Hanbury, Hazard, C., Davis, J., and Allen. L.R. (1964). *Nature* **201**, 1111–1112.
Brown, R. Hanbury, and Twiss, R.Q. (1954). *Phil. Mag.* **45**, 663–682.
Brown, R. Hanbury and Twiss, R.Q. (1956). *Nature* **177**, 27–29.
Brown, R. Hanbury and Twiss, R.Q. (1957a). *Proc. Roy. Soc.* A **242**, 300–324.
Brown, R. Hanbury and Twiss, R.Q. (1957b). *Proc. Roy. Soc.* A **243**, 291–319.
Brown, R. Hanbury and Twiss, R.Q. (1958a). *Proc. Roy. Soc.* A **248**, 199–221.
Brown, R. Hanbury and Twiss, R.Q. (1958b). *Proc. Roy. Soc.* A **248**, 222–237.
Bruce, C.F., and Hill, R.M. (1961). *Aust. J. Phys.* **14**, 64–88.
Bruning, J.H. (1978). *In* "Optical Shop Testing" (D. Malacara, ed.), pp. 409–437. John Wiley, New York.
Bruning, J.H., Herriott, D.R., Gallagher, J.E., Rosenfeld, D.P., White, A.D., and Brangaccio, D.J. (1974). *Appl. Opt.* **13**, 2693–2703.
Bryngdahl, O. (1965). *In* "Progress in Optics" (E. Wolf, ed.), Vol. IV, pp. 37–83. North-Holland, Amsterdam.
Bryngdahl, O. (1969). *J. Opt. Soc. Am.* **59**, 142–146.
Bryngdahl, O., and Lohmann, A. (1968). *J. Opt. Soc. Am.* **58**, 1325–1334.
Bucaro, J.A., Dardy, H.D., and Carome, E.F. (1977). *Appl. Opt.* **16**, 1761–1762.
Bünnagel, R. von, (1956). *Z. Angew. Phys.* **8**, 342–350.
Bünnagel, R. von, (1965). *Z. Instrumkde.* **73**, 214–215.
Burch, J.M. (1953). *Nature* **171**, 889–890.
Burch, J.M. (1965). *Prodn. Engnr.* **44**, 431–442.
Burch, J.M. (1972). *In* "Optical Instruments and Techniques" (J. Home-Dickson, ed.), pp. 213–229. Oriel Press, Newcastle-upon-Tyne.
Butter, C.D., and Hocker, G.B. (1978). *Appl. Opt.* **17**, 2867–2869.
Butters, J.N., and Leendertz, J.A. (1971a). *J. Phys. E: Sci. Instrum.* **4**, 277–279.
Butters, J.N., and Leendertz, J.A. (1971b). *Opt. Laser Technol.* **3**, 26–30.
Byer, R.L., Paul, J., and Duncan, M.D. (1977). *In* "Laser Spectroscopy III" (J.L. Hall and J.L. Carlsten, eds), pp. 414–416. Springer-Verlag, Berlin.
Cagnet, M. (1954). *Rev. Opt.* **33**, 1–25, 113–124, 229–241.
Candler, C. (1951). "Modern Interferometers". Hilger and Watts, London.
Carré, P. (1966). *Metrologia* **2**, 13–23.
Caulfield, H.J. (1979). "Handbook of Optical Holography". Academic Press, New York.
Cha, S., and Vest, C.M. (1979). *Opt. Lett.* **4**, 311–313.
Cha, S., and Vest, C.M. (1981). *Appl. Opt.* **20**, 2787–2794.
Chabbal, R. (1958). *Rev. Opt.* **37**, 49–103, 366–370, 501–551.
Chakraborty, A.K. (1970). *Ind. J. Pure and Appl. Phys.* **8**, 814–816.
Chakraborty, A.K. (1973). *Nouv. Rev. Optique* **4**, 331–335.
Chamberlain, J. (1971). *Infrared Physics* **11**, 25–55.
Chamberlain, J. (1979). "The Principles of Interferometric Spectroscopy". Wiley, Chichester.
Chamberlain, J., and Gebbie, H.A. (1971). *Infrared Physics* **11**, 56–73.
Chang, B.J., and Leonard, C.D. (1979). *Appl. Opt.* **18**, 2407–2417.
Ciddor, P.E. (1973). *Aust. J. Phys.* **26**, 783–796.
Ciddor, P.E., and Bruce, C.F. (1967). *Metrologia* **3**, 109–118.
Ciddor, P.E., and Duffy, R.M. (1983). *J. Phys. E: Sci. Instrum.* **16**, 1223–1227.

Clapham, P.B., Downs, M.J., and King, R.J. (1969). *Appl. Opt.* **8**, 1965–1974.

Clark, R.J., Hause, C.D. and Bennett, G.S. (1953). *J. Opt. Soc. Am.* **43**, 408–409.

Close, D.H. (1975). *Opt. Eng.* **14**, 408–419.

Clunie, D.M. and Rock, N.H. (1964). *J. Sci. Instrum.* **41**, 489–492.

Cochran, W.T., Cooley, J.W., Favin, D.L., Helms, H.D., Kaenel, R.A., Lang, W.W., Maling, G.C., Jr., Nelson, D.E., Rader, C.M., and Welch, P.D. (1967). *Proc. IEEE* **55**, 1664–1674.

Collier, R.J., Burckhardt, C.B., and Lin, L.H. (1971). "Optical Holography", Academic Press, New York.

Collier, R.J., Doherty, E.T., and Pennington, K.S. (1965). *Appl. Phys. Lett.* **7**, 223–225.

Collins, R.J., Nelson, D.F., Schawlow, A.L., Bond, W., Garrett, C.G.B., and Kaiser, W. (1960). *Phys. Rev. Lett.* **5**, 303–305.

Connes, J. (1961). *Rev. Opt.* **40**, 45–79, 116–140, 171–190, 231–265.

Connes, J., and Connes, P. (1966). *J. Opt. Soc. Am.* **56**, 896–910.

Connes, J., and Nozal, V. (1961). *J. Phys. Radium.* **22**, 359–366.

Connes, P. (1956). *Rev. Opt.* **35**, 37–43.

Connes, P. (1958). *J. Phys. Radium.* **19**, 262–269.

Cookson, T.J., Butters, J.N., and Pollard, H.G. (1978). *Opt. Laser Technol.* **10**, 119–124.

Cooley, J.W., and Tukey, J.W. (1965). *Math Comput.* **19**, 297–301.

Cox, J.T., Hass, G., and Thelen, A. (1962). *J. Opt. Soc. Am.* **52**, 965–969.

Crane, R. (1969). *Appl. Opt.* **8**, 538–542.

Dahlquist, J.A., Peterson, D.G., and Culshaw, W. (1966). *Appl. Phys. Lett.* **9**, 181–183.

Dainty, J.C. (1973). *Opt. Commun.* **7**, 129–134.

Dainty, J.C. (1975). "Laser Speckle and Related Phenomena". Springer-Verlag, Berlin.

Dainty, J.C. (1976). *In* "Progress in Optics" (E. Wolf, ed), Vol. XIV, pp. 3–46. North-Holland, Amsterdam.

Dallas, W.J. (1980). *In* "The Computer in Optical Research" (B.R. Frieden, ed.), pp. 291–366. Springer-Verlag, Berlin.

Dändliker, R. (1980). *In* "Progress in Optics" (E. Wolf. ed.), Vol. XVII, pp. 1–84. North-Holland, Amsterdam.

Dändliker, R., Ineichen, B., and Mottier, F.M. (1973). *Opt. Commun.* **9**, 412–416.

Dandridge, A., Miles, R.O., and Giallorenzi, T.G. (1980). *Electron. Lett.* **16**, 948–949.

Dandridge, A., and Tveten, A.B. (1981). *Appl. Opt.* **20**, 2337–2339.

Davis, J. (1979). *In* "High Angular Resolution Stellar Interferometry" (J. Davis and W.J. Tango, eds), I.A.U. Colloquium No. 50 pp. 14.1 – 14.12. Chatterton Astronomy Department, University of Sydney.

Davis, J. (1984). *Proc. International Symposium on Measurement and Processing for Indirect Imaging, Sydney, 1983* (J. Roberts, ed.), pp. 125–141. Cambridge University Press, Cambridge.

Denby, D., and Leendertz, J.A. (1974). *J. Strain Anal.* **9**, 17–25.

Denisyuk, Yu. N. (1962). *Soviet Phys. Doklady* **7**, 543–545.

Dörband, D. (1982). *Optik* **60**, 161–174.

Drever, R.W.P., Hough, J., Munley, A.J., Lee, S.A., Spero, R., Whitcomb, S.E., Ward, H., Ford, G.M., Hereld, M., Robertson, N.A., Kerr, I., Pugh, J.R., Newton, G.P., Meers, B., Brooks III, E.D., Gursel, Y. (1981). *In* "Laser Spectroscopy V" (A.R.W. McKellar, T. Oka and B.P. Stoicheff, eds), pp. 33–40. Springer-Verlag, Berlin.

Dubas, M., and Schumann, W. (1974). *Opt. Acta.* **21**, 547–562.

Dubas, M., and Schumann, W. (1977). *Opt. Acta.* **24**, 1193–1209.

Duffy, D.E. (1972). *Appl. Opt.* **11**, 1778–1781.

Dufour, C. (1951). *Ann. Phys.* **6**, 5–107.

Dukes, J.N., and Gordon, G.B. (1970). *Hewlett-Packard Journal,* **21**, (No. 12), 2–8.

Dyson, J. (1957). *J. Opt. Soc. Am.* **47**, 386–390.

Dyson, J. (1963). *J. Opt. Soc. Am.* **53**, 690-694.
Dyson, J., Flude, M.J.C., Middleton, S.P., and Palmer, E.W. (1972). *In* "Optical Instruments and Techniques" (J. Home Dickson, ed.), pp 93-103. Oriel Press, Newcastle-upon-Tyne.
Edlén, B. (1966). *Metrologia,* **2,** 71-80.
Ehn, D.C., and Nisenson, P. (1975). *J. Opt. Soc. Am.* **65,** 1196.
Ek. L., and Molin, N.E. (1971). *Opt. Commun.* **2,** 419-424.
Ennos, A.E. (1968). *J. Phys. E: Sci. Instrum.* **1,** 731-734.
Ennos, A.E. (1978) *In* "Progress in Optics" (E. Wolf, ed.), Vol. XVI, pp. 235-288. North-Holland, Amsterdam.
Epstein, L.I. (1955). *J. Opt. Soc. Am.* **45,** 360-362.
Erf, R.K. (1974). "Holographic Non-destructive Testing". Academic Press, New York.
Erf, R.K. (1978). "Speckle Metrology". Academic Press, New York.
Evenson, K.M., Day, G.W., Wells, J.S., and Mullen, L.O. (1972). *Appl. Phys. Lett.* **20**, 133-134.
Evenson, K.M., Jennings, D.A., and Petersen, F.R. (1981). *J. Phys. Colloq. (France).* **42 (C-8),** 473-483.
Evenson, K.M., Wells, J.S., Petersen, F.R., Danielson, B.L., and Day, G.W. (1973). *Appl. Phys. Lett.* **22**, 192-195.
Ezekiel, S., and Arditty, H.J. (eds) (1982). "Fiber-Optic Rotation Sensors and Related Technologies". Springer-Verlag, New York.
Ezekiel, S., and Balsamo, S.R. (1977). *Appl. Phys. Lett.* **30**, 478-480.
Ezekiel, S., and Knausenberger, G.E., (1978). *Proc. SPIE, Vol. 157.* "Laser Inertial Rotation Sensors", SPIE, Bellingham.
Faulde, M., Fercher, A.F., Torge, R., and Wilson, R.N. (1973). *Opt. Commun.* **7**, 363-365.
Fellgett, P.B. (1951). Thesis, University of Cambridge, England.
Fellgett, P. (1958). *J. Phys. Radium.* **19**, 237-240.
Ferguson, J.B., and Morris, R.H. (1978). *Appl. Opt.* **17**, 2924-2929.
Fienup, J.R. (1978). *Opt. Lett.* **3**, 27-29.
Fischer, A., Kullmer, R., and Demtroder, W. (1981). *Opt. Comm.* **39**, 277-282.
Fork. R.L., Herriott, D.R., and Kogelnik, H. (1964). *Appl. Opt.* **3**, 1471-1484.
Forman, M.L. (1966). *J. Opt. Soc. Am.* **56**, 978-979.
Forman, M., Steel, W.H., and Vanasse, G.A. (1966). *J. Opt. Soc. Am.* **56,** 59-63.
Françon, M. (1957). *J. Opt. Soc. Am.,* **47**, 528-535.
Françon, M. (1961). "Progress in Microscopy". Pergamon, Oxford.
Françon, M., and Mallick, S. (1971). "Polarization Interferometers: Applications in Microscopy and Macroscopy". Wiley-Interscience, London.
Frantz, L.M., Sawchuk, A.A., and von der Ohe, W. (1979). *Appl. Opt.* **18**, 3301-3306.
Freed, C., and Javan, A. (1970). *Appl. Phys. Lett.* **17**, 53-56.
Freniere, E.R., Toler, O.E., and Race, R. (1981). *Opt. Eng.* **20**, 253-255.
Friesem, A.A., and Levy, U. (1976). *Appl. Opt.* **15**, 3009-3020.
Gabor, D. (1948). *Nature* **161**, 777-778.
Gabor, D. (1949). *Proc. Roy. Soc.* **A 197**, 454-487.
Gabor, D. (1951). *Proc. Phys. Soc.* **B 64**, 449-469.
Gardner, J.L. (1983). *Opt. Lett.* **8**, 91-93.
Gates, J.W. (1955). *Nature* **176**, 359-360.
Gates, J.W.C. (1968). *Nature* **220**, 473-474.
Gates, J.W.C., Hall, R.G.N., and Ross, I.N. (1972). *Opt. Laser Tech.* **4**, 72-75.
Gezari, D.Y., Labeyrie, A., and Stachnik, V. (1972). *Astrophys. J.* **173**, L1-L5.
Giallorenzi, T.G., Bucaro, J.A., Dandridge, A., Sigel, G.H., Jr., Cole, J.H., Rashleigh, S.C., and Priest, R.G. (1982). *IEEE J. Quant. Electron.* **QE-18**, 626-665.
Gillard, C.W., and Buholz, N.E. (1983). *Opt. Eng.* **22**, 348-353.
Gillard, C.W., Buholz, N.E., and Ridder, D.W. (1981). *Opt. Eng.* **20**, 129-134.

Gilliland, K.E., Cook. H.D., Mielenz, K.D., and Stephens, R.B. (1966). *Metrologia* **2**, 95–98.

Goldberg, J.L. (1975). *Japan J. Appl. Phys.* **14, Suppl. 14-1**, 253–258.

Goldfischer, L.I. (1965). *J. Opt.Soc. Am.* **55**, 247–253.

Goodman, J.W. (1968). "Introduction to Fourier Optics", McGraw-Hill, New York.

Goodman, J.W. (1975). *In* "Laser Speckle and Related Phenomena" (J.C. Dainty, ed.), pp. 9–75. Springer-Verlag, Berlin.

Gordon, S.K., and Jacobs, S.F. (1974). *Appl. Opt.* **13**, 231.

Gough, P.T., and Bates, R.H.T., (1974). *Opt. Acta* **21**, 243–254.

Gregory, D.A. (1976a). *In* "The Engineering Uses of Coherent Optics" (E.R. Robertson, ed.), pp. 263–282. Cambridge University Press, Cambridge.

Gregory, D.A. (1976b). *Opt. Laser Tech.* **8**, 201–213.

Greguss, P. (1975). "Holography in Medicine". IPC Press, London.

Guelachvili, G. (1978). *Appl. Opt.* **17**, 1322–1326.

Haines, K.A., and Hildebrand, B.P. (1965). *Phys. Lett.* **19**, 10–11.

Hall, J.L. (1968). *IEEE J. Quantum Electron.* **QE-4**, 638–641.

Hall, J.L., and Lee, S.A. (1976). *Appl. Phys. Lett.* **29**, 367–369.

Hall, R.N., Fenner, G.E., Kingley, J.D., Soltys, T.J., and Carlson, R.O. (1962). *Phys. Rev. Lett.* **9**, 366–368.

Hanel, R., and Kunde, V.G. (1975). *Space Sci. Rev.* **18**, 201–256.

Hanes, G.R. (1963). *Appl. Opt.* **2**, 465–470.

Hanes, G.R., Baird, K.M., and De Remigis, J. (1973). *Appl. Opt.* **12**, 1600–1605.

Hanes, G.R., and Dahlstrom, C.E. (1969). *Appl. Phys. Lett.* **14**, 362–364.

Hansen, G. (1955). *Optik* **12**, 5–16.

Hansen, G., and Kinder, W. (1958). *Optik* **15**, 560–564.

Hariharan, P. (1969a). *Appl. Opt.* **8**, 1925–1926.

Hariharan, P. (1969b). *J. Opt. Soc. Am.* **59**, 1384.

Hariharan, P. (1972). *Opt. Acta* **19**, 791–793.

Hariharan, P. (1973). *Appl. Opt.* **12**, 143–146.

Hariharan, P. (1975a). *Opt. Eng.* **14**, 257–258.

Hariharan, P. (1975b). *Appl. Opt.* **14**, 1056–1057.

Hariharan, P. (1975c). *Appl. Opt.* **14**, 2319–2321.

Hariharan, P. (1975d). *Appl. Opt.* **14**, 2563.

Hariharan, P. (1977). *Opt. Acta* **24**, 979–987.

Hariharan, P. (1978). *Opt. Commun.* **26**, 325–326.

Hariharan, P. (1980). *Opt. Eng.* **19**, 636–641.

Hariharan, P. (1984). "Optical Holography: Theory, Techniques and Applications". Cambridge University Press, Cambridge.

Hariharan, P., and Hegedus, Z.S. (1973). *Opt. Commun.* **9**, 152–155.

Hariharan, P., and Hegedus, Z.S. (1974). *Opt. Commun.* **14**, 148–151.

Hariharan, P., and Hegedus, Z.S. (1976). *Appl. Opt.* **15**, 848–849.

Hariharan, P., Oreb, B.F., and Brown, N. (1982). *Opt. Commun.* **41**, 393–396.

Hariharan, P., Oreb, B.F., and Brown, N. (1983). *Appl. Opt.* **22**, 876–880.

Hariharan, P., Oreb, B.F., and Leistner, A.J. (1984). *Opt. Eng.* **23**, 294–297.

Hariharan, P., and Ramprasad, B.S. (1973a). *J. Phys. E: Sci. Instrum.* **6**, 173–175.

Hariharan, P., and Ramprasad, B.S. (1973b). *J. Phys. E: Sci. Instrum.* **6**, 699–701.

Hariharan, P., and Sen, D. (1959). *J. Sci. Instrum.* **36**, 70–72.

Hariharan, P., and Sen, D. (1960a). *J. Opt. Soc. Am.* **50**, 1026–1027.

Hariharan, P., and Sen, D. (1960b). *J. Opt. Soc. Am.* **50**, 357–361.

Hariharan, P., and Sen, D. (1960c). *J. Opt. Soc. Am.* **50**, 999–1001.

Hariharan, P., and Sen, D. (1960d). *J. Sci. Instrum.* **37**, 374–376.

Hariharan, P., and Sen, D. (1960e). *Proc. Phys. Soc.* **75**, 434–438.
Hariharan, P., and Sen, D. (1961a). *J. Opt. Soc. Am.* **51**, 1307.
Hariharan, P., and Sen, D. (1961b). *J. Opt. Soc. Am.* **51**, 400–404.
Hariharan, P., and Sen, D. (1961c). *J. Opt. Soc. Am* **51**, 1212–1218.
Hariharan, P., and Sen, D. (1961d). *Proc. Phys. Soc.* **77**, 328–334.
Hariharan, P., and Sen, D. (1961e). *J. Sci. Instrum.* **38**, 428–432.
Hariharan, P., and Sen, D. (1961f). *J. Opt. Soc. Am.* **51**, 398–399.
Hariharan, P., and Singh, R.G. (1959a). *J. Sci. Instrum.* **36**, 323–324.
Hariharan, P., and Singh, R.G. (1959b). *J. Opt. Soc. Am.* **49**, 732–733.
Hariharan, P., Steel, W.H., and Wyant, J.C. (1974). *Opt. Commun.* **11**, 317–320.
Harris, S.E., Ammann, E.O., and Chang, I.C. (1964). *J. Opt. Soc. Am.* **54**, 1267–1279.
Hayashi, I., Panish, M.B., and Foy, P.W. (1969). *IEEE J. Quant. Electron.* **QE-5**, 211–212.
Heavens, O.S., and Liddell, H.M. (1966). *Appl. Opt.* **5**, 373–376.
Heintze, L.R. (1967). *Appl. Opt.* **6**, 1924–1929.
Hercher, M. (1968). *Appl. Opt.* **7**, 951–966.
Hercher, M. (1969). *Appl. Opt.* **8**, 1103–1106.
Herpin, A. (1947). *C.R. Acad. Sci. Paris.* **225**, 182–183.
Herriot, D.R. (1962). *J. Opt. Soc. Amer.* **52**, 31–37.
Herriott, D.R. (1963). *Appl. Opt.* **2**, 865–866.
Hicks, T.R., Reay, N.K., and Scaddan, R.J. (1974). *J. Phys. E; Sci. Instrum.* **7**, 27–30.
Hildebrand, B.P., and Haines, K.A. (1966). *Phys. Lett.* **21**, 422–423.
Hildebrand, B.P., and Haines, K.A. (1967). *J. Opt. Soc. Am.* **57**, 155–162.
Hill, R.M., and Bruce, C.F. (1962). *Aust. J. Phys.* **15**, 194–222.
Hill, R.M., and Bruce, C.F. (1963). *Aust. J. Phys.* **16**, 282–285.
Hocker, G.B. (1979). *Appl. Opt.* **18**, 1445–1448.
Hodgkinson, I.J., and Vukusic, J.I. (1978). *Appl. Opt.* **17**, 1944–1948.
Høgmoen, K., and Pedersen, H.M. (1977). *J. Opt. Soc. Am.* **67**, 1578–1583.
Holonyak, Jr., N., and Bevacqua, S.F. (1962). *Appl. Phys. Lett.* **1**, 82–83.
Holtom, G., and Teschke, O. (1974). *IEEE J. Quant. Electron.* **QE-10**, 577–579.
Hopf. F.A., Tomita, A., and Al-Jumaily, G. (1980). *Opt. Lett.* **5**, 386–388.
Hopkins, H.H. (1951). *Proc. Roy. Soc.* **A 208**, 263–277.
Hopkins, H.H. (1953). *Proc. Roy. Soc.* **A 217**, 408–432.
Hopkins, H.H. (1955). *Opt. Acta* **2**, 23–29.
Hopkinson, G.R. (1978). *J. Optics (Paris)* **9**, 151–155.
Houston, J.B., Jr., Buccini, C.J., and O'Neill, P.K. (1967). *Appl. Opt.* **6**, 1237–1242.
Huignard, J.P., Herriau, J.P. Aubourg, P., and Spitz, E. (1979). *Opt. Lett.* **4**, 21–23.
Huignard, J.P., and Micheron, F. (1976). *Appl. Phys. Lett.* **29**, 591–593.
Hung, Y.Y., Hu, C.P., Henley, D.R., and Taylor, C.E. (1973). *Opt. Commun.* **8**, 48–51.
Hung, Y.Y., and Liang, C.Y. (1979). *Appl. Opt.* **18**, 1046–1051.
Hung, Y.Y., Rowlands, R.E., and Daniel, I.M. (1975). *Appl. Opt.* **14**, 618–622.
Hung, Y.Y., and Taylor, C.E. (1974). *Proc. SPIE* **41**, 169–175.
Hutley, M.C. (1976). *J. Phys. E: Sci. Instrum.* **9**, 513–520.
Hutley, M.C. (1982). "Diffraction Gratings". Academic Press, London.
Ja, Y.H. (1982). *Opt. Quant. Electron.* **14**, 367–369.
Jackson, D.A., Dandridge, A., and Sheen, S.K. (1980a). *Opt. Lett.* **5**, 139–141.
Jackson, D.A., Priest, R., Dandridge, A., and Tveten, A.B. (1980b). *Appl. Opt.* **19**, 2926–2929.
Jacobsson, R. (1965). *In* "Progress in Optics" (E. Wolf, ed.), Vol. V, pp. 249–286. North-Holland, Amsterdam.
Jacquinot, P. (1960). *Rep. Progr. Phys.* **23**, 267–312.
Jacquinot, P., and Dufour, C. (1948). *J. Rech. C.N.R.S.* **6**, 91–103.

Jacquinot, P., and Roizen-Dossier, B. (1964). *In* "Progress in Optics" (E. Wolf, ed), Vol. III, pp. 29–186. North-Holland, Amsterdam.

Jahns, J., and Lohmann, A.W. (1979). *Opt. Commun.* **28**, 263–267.

Jarrett, S.M., and Young, J.F. (1979). *Opt. Lett.* **4**, 176–178.

Jaseja, T.S., Javan, A., Murray, J., and Townes, C.H. (1964). *Phys. Rev.* **133A**, 1221–1225.

Javan, A., Ballik, E.A., and Bond, W.L. (1962). *J. Opt. Soc. Am.* **52**, 96–98.

Javan, A., Bennett, W.R., and Herriott, D.R. (1961). *Phys. Rev. Lett.* **6**, 106–110.

Jerrard, H.G. (1982) *Opt. and Laser Tech.* **14**, 309–319.

Johnson, G.W., Leiner, D.C., and Moore, D.T., (1977). *Proc SPIE* **126**, 152–160.

Jones, F.E. (1981). *J. Res. NBS.* **86**, 27–32.

Jones, R., and Butters, J.N. (1975). *J. Phys. E: Sci. Instrum.* **8**, 231–234.

Jones, R.C. (1941). *J. Opt. Soc. Am.* **31**, 488–493.

Juncar, P., and Pinard, J. (1975). *Opt. Commun.* **14**, 438–441.

Juncar, P., and Pinard, J. (1982). *Rev. Sci. Instrum.* **53**, 939–948.

Kajimura, T., Koroda, T., Yamashita, S., Nakamura, M., and Umeda, J. (1979). *Appl. Opt.* **18**, 1812–1815.

Kelsall, D. (1959). *Proc. Phys. Soc.* **73**, 465–479.

Kennedy, R.J. (1926). *Proc. Nat. Acad. Sci. Wash.* **12**, 621–629.

Kingslake, R. (1925–26). *Trans. Opt. Soc.* **27**, 94–105.

Kinosita, K. (1953). *J. Phys. Soc. Japan.* **8**, 219–225.

Knight, D.J.E., Edwards, G.J., Pearce, P.R., and Cross, N.R. (1980). *IEEE Trans. Instrum. Measurement.* **IM-29**, 257–264.

Knox, J.D., and Pao, Y. (1970). *Appl. Phys. Lett.* **16**, 129–131.

Knox, K.T., and Thompson, B.J. (1974). *Astrophys. J.* **193** L45–L48.

Koechner, W. (1976). "Solid-State Laser Engineering". Springer-Verlag, New York.

Kogelnik, H. (1965). *Bell Syst. Tech. J.* **44**, 2451–2455.

Kogelnik, H. (1967). *In* "Proceedings of the Symposium on Modern Optics", pp. 605–617. Polytechnic Press, Brooklyn.

Kogelnik, H. (1969). *Bell Syst. Tech. J.* **48**, 2909–2947.

Kogelnik, H. and Li, T. (1966). *Appl. Opt.* **5**, 1550–1567.

Koppelmann, G., and Krebs, K. (1961a). *Optik* **18**, 349–357.

Koppelmann, G., and Krebs, K. (1961b). *Optik* **18**, 358–372.

Kowalski, F.V., Hawkins, R.T., and Schawlow, A.L. (1976). *J. Opt. Soc. Am.* **66**, 965–966.

Kowalski, F.V., Teets, R.E., Demtröder, W., and Schawlow, A.L. (1978). *J. Opt. Soc. Am.* **68**, 1611–1613.

Kressel, H., and Nelson, H. (1969). *RCA Rev.* **30**, 106–113.

Kristal, R., and Peterson, R.W. (1976). *Rev. Sci. Instrum.* **47**, 1357–1359.

Krug, W., Rienitz, J., and Schulz, G. (1964). "Contributions to Interference Microscopy". Hilger and Watts, London.

Kubota, H. (1950). *J. Opt. Soc. Am.* **40**, 146–149.

Kubota, H. (1961). *In* "Progress in Optics" (E. Wolf, ed.), Vol. I, pp. 213–251. North-Holland, Amsterdam.

Kwon, O., Wyant, J.C., and Hayslett, C.R. (1980). *Appl. Opt.* **19**, 1862–1869.

Kyuma, K., Tai, S., and Nunoshita, M. (1982). *Opt. and Lasers Eng.* **3**, 155–182.

Labeyrie, A. (1970). *Astron. and Astrophys.* **6**, 85–87.

Labeyrie, A. (1974). *Nouv. Rev. Optique,* **5**, 141–151.

Labeyrie, A. (1976). *In* "Progress in Optics" (E. Wolf, ed.), Vol. XIV, pp.49–87. North-Holland, Amsterdam.

Labeyrie, A. (1978). *Ann. Rev. Astron. Astrophys.* **16**, 77–102.

Labeyrie, A., Bonneau, D., Stachnik, R.V., and Gezari, D.Y. (1974). *Astrophys. J.* **194**, L147–L151.

Lavan, M.J., Cadwallender, W.K., De Young, T.F., and Van Damme, G.E. (1976). *Appl. Opt.* **15**, 2627–2628.

Layer, H.P. (1980). *IEEE Trans. Instrum. Measurement.* **IM-29**, 358–361.

Lee, P.H., and Skolnick, M.L. (1967). *Appl. Phys. Lett.* **10**, 303–305.

Lee, W.H. (1974). *Appl. Opt.* **13**, 1677–1682.

Lee, W.H. (1978). *In* "Progress in Optics" (E. Wolf, ed.), Vol. XVI, pp. 121–232. North-Holland, Amsterdam.

Lee, W.H. (1979). *Appl. Opt.* **18**, 3661–3669.

Leeb, W.R., Schiffner, G., and Scheiterer, E.. (1979). *Appl. Opt.* **18**, 1293–1295.

Leendertz, J.A. (1970). *J. Phys. E: Sci. Instrum.* **3**, 214–218.

Leendertz, J.A., and Butters, J.N. (1973). *J. Phys. E: Sci. Instrum.* **6**, 1107–1110.

Leith, E.N., and Upatnieks, J. (1962). *J. Opt. Soc. Am.* **52**, 1123–1130.

Leith, E.N., and Upatnieks, J. (1963). *J. Opt. Soc. Am.* **53**, 1377–1381.

Leith, E.N., and Upatnieks, J. (1964). *J. Opt. Soc. Am.* **54**, 1295–1301.

Lenouvel, L., and Lenouvel, F. (1938). *Rev. Opt.* **17**, 350–361.

Leonhardt, K. (1972). *Optik* **35**, 509–523.

Leonhardt, K., (1974). *Opt. Commun.* **11**, 312–316.

Leonhardt, K. (1981). "Optische Interferenzen". Wissenschaftliche Vergesselschaft, Stuttgart.

Levenson, M. (1981). *Holosphere,* **10 (7&8)**, 7.

Liewer, K.M. (1979). *In* "High Angular Resolution Stellar Interferometry" (J. Davis and W.J. Tango, eds.), I.A.U. Colloquium No. 50, pp. 8.1 – 8.14. Chatterton Astronomy Department, University of Sydney.

Lin, L.H., and Beauchamp, H.L. (1970). *Appl. Opt.* **9**, 2088–2092.

Lin, S.C., and Giallorenzi, T.G. (1979). *Appl. Opt.* **18**, 915–931.

Lindsay, S.M., and Shepherd, I.W. (1977). *J. Phys. E: Sci. Instrum.* **10**, 150–154.

Linnik, V.P. (1942). *Comptes Rendus (Doklady) URSS,* **35**, 16–19.

Liu, C.Y.C., and Lohmann, A.W. (1973). *Opt. Commun.* **8**, 372–377.

Liu, L.S., and Klinger, J.H. (1979). *Proc. SPIE* **192**, 17–26.

Lohmann, A.W., and Paris, D.P. (1967). *Appl. Opt.* **6**, 1739–1748.

Lohmann, A.W., and Silva, D.E. (1971). *Opt. Commun.* **2**, 413–415.

Lohmann, A.W., and Weigelt, G.P. (1979). *In* "Proceedings of the European Workshop on Astronomical Uses of the Space Telescope" (F. Macchetto, F. Pacini and M. Tarenghi, eds), pp. 353–361. ESO, Geneva.

Lohmann, A.W., Weigelt, G., and Wirnitzer, B. (1983). *Appl. Opt.* **22**, 4028–4037.

Løkberg, O.J. (1979). *Appl. Opt.* **18**, 2377–2384.

Løkberg, O.J. and Høgmoen, K. (1976a). *J. Phys. E: Sci. Instrum.* **9**, 847–851.

Løkberg, O.J., and Høgmoen, K. (1976b). *Appl. Opt.* **15**, 2701–2704.

Løkberg, O.J., Høgmoen, K., and Holje, O.M. (1979). *Appl. Opt.* **18**, 763–765.

Løkberg, O.J., Holje, O.M., and Pedersen, H.M. (1976). *Opt. Laser Technol.* **8**, 17–20.

Loomis, J.S. (1980). *Opt. Eng.* **19**, 679–685.

Lowenthal, S., and Arsenault, H.H. (1970). *J. Opt. Soc. Am.* **60**, 1478–1483.

Lowenthal, S., Serres, J., and Froehly, C. (1969). *C.R. Acad. Sci. Paris,* **268B**, 841–844.

Luc, P., and Gerstenkorn, S. (1978). *Appl. Opt.* **17**, 1327–1331.

Lynds, G.R., Worden, S.P., and Harvey, J.W. (1976). *Astrophys. J.* **207**, 174–180.

Lyot, B. (1944). *Ann. Astrophys.* **7**, 31–79.

Macek, W., and Davis, D. (1963). *Appl. Phys. Lett.* **2**, 67–68.

MacGovern, A.J., and Wyant, J.C. (1971). *Appl. Opt.* **10**, 619–624.

Mack, J.E., McNutt, D.P., Rossler, F.L., and Chabbal, R. (1963). *Appl. Opt.* **2**, 873–885.

Macleod, H.A. (1969). "Thin-Film Optical Filters". Adam Hilger, London.

Macleod, H.A. (1972). *Opt. Acta,* **19**, 1–28.

Macovski, A., Ramsey, S.D., and Schaefer, L.F. (1971). *Appl. Opt.* **10**, 2722–2727.

Mahlein, H.F. (1974). *Opt. Acta.* **21**, 577-583.
Maiman, T.H. (1960). *Nature* **187**, 493-494.
Maischberger, K., Rudiger, A., Schilling, R., Schnupp, L., Winkler, W., and Billing, H. (1981). *In* "Laser Spectroscopy V" (A.R.W. McKellar, T. Oka and B.P. Stoicheff, eds), pp. 25-32. Springer-Verlag, Berlin.
Malacara, D. (1974). *Appl. Opt.* **13**, 1781-1784.
Malacara, D. (1978a). *In* "Optical Shop Testing" (D. Malacara, ed.), pp. 47-49. John Wiley, New York.
Malacara, D. (1978b). *In* "Optical Shop Testing" (D. Malacara, ed.), pp. 149-178. John Wiley, New York.
Malacara, D., and Mendez, M. (1968). *Opt. Acta.* **15**, 59-63.
Malyshev, Yu. M., Ovchimikov, S.N., Rastorguev, Yu. G., Tatarenkov, V.M., and Titov, A.N. (1980). *Sov. J. Quant. Electron.* **10**, 376-377.
Mandel, L. (1963). *In* "Progress in Optics" (E. Wolf, ed.), Vol II, pp. 181-248. North-Holland, Amsterdam.
Mandel, L., and Wolf, E. (1965). *Rev. Mod. Phys.* **37**, 231-287.
Maréchal, A., Lostis, P., and Simon, J. (1967). *In* "Advanced Optical Techniques" (A.C.S. van Heel, ed.), pp. 435-446. North-Holland, Amsterdam.
Marrakchi, A., Huignard, J.P., and Herriau, J.P. (1980). *Opt. Commun.* **34**, 15-18.
Massie, N. A. (1980). *Appl Opt.* **19**, 154-160.
Massie, N.A., Nelson, R.D., and Holly, S. (1979). *Appl. Opt.* **18**, 1797-1803.
Mertz, L. (1965). "Transformations in Optics". John Wiley, New York.
Michelson, A. A., and Pease, F.G. (1921). *Astrophys. J.* **53**, 249-259.
Monchalin, J.P., Kelly, M.J., Thomas, J.E., Kurnit, N.A., Szöke, A., Zernike, F., Lee, P.H., and Javan, A. (1981). *Appl. Opt.* **20**, 736-757.
Montgomery, A.J. (1964). *J. Opt. Soc. Amer.* **54**, 191-198.
Morris, R.H. Ferguson, J.B., and Warniak, J.S. (1975). *Appl. Opt.* **14**, 2808.
Mottier, F.M. (1979). *Opt. Eng.* **18**, 464-468.
Munnerlyn, C.R. (1969). *Appl. Opt.* **8**, 827-829.
Munnerlyn, C.R., and Latta, M. (1968). *Appl Opt.* **7**, 1858-1859.
Murty, M.V.R.K. (1963). *J. Opt. Soc. Am.* **53**, 568-570.
Murty, M.V.R.K. (1964). *Appl. Opt.* **3**, 531-534.
Murty, M.V.R.K. (1978). *In* "Optical Shop Testing" (D. Malacara, ed.). pp. 105-148. John Wiley, New York.
Murty, M.V.R.K., and Hagerott, E.C. (1966). *Appl. Opt.* **5**, 615-619.
Murty, M.V.R.K., and Shukla, R.P. (1976). *Opt. Eng.* **15**, 461-463.
Musset, A., and Thelen, A. (1970). *In* "Progress in Optics" (E. Wolf, ed.), Vol. VIII, pp. 203-237. North-Holland, Amsterdam.
Nakadate, S., Magome, N., Honda, T., and Tsujiuchi, J. (1981). *Opt. Eng.* **20**, 246-252.
Nakadate, S., Yatagai, T., and Saito, H. (1980a). *Appl. Opt.* **19**, 1879-1883.
Nakadate, S., Yatagai, T., and Saito, H. (1980b). *Appl. Opt.* **19**, 4241-4246.
Nathan, M.I., Dumke, W.P., Burns, G., Dill, F.H., and Lasher, G. (1962). *Appl. Phys. Lett.* **1**, 62-64.
Netterfield, R.P. (1977). *Opt. Acta.* **24**, 69-79.
Neumann, D.B., Jacobson, C.F., and Brown, G.M. (1970). *Appl. Opt.* **9**, 1357-1368.
Öhman, Y. (1938). *Nature* **141**, 157-158, 291.
Ohtsuka, Y., and Sasaki, I. (1974). *Opt. Commun.* **10**, 362-365.
Ostrovskaya, G.V., and Ostrovskii, Yu. I. (1971). *Sov. Phys. Tech. Phys.* **15**, 1890-1892.
Panarella, E. (1973). *J. Phys. E: Sci. Instrum* **6**, 523.
Pancharatnam, S. (1963). *Proc. Ind. Acad. Sci.* **57**, 231-243.

Panish, M.B., Hayashi, I., and Sumski, S. (1970). *Appl. Phys. Lett.* **16**, 326–327.
Papoulis, A. (1965). "Probability, Random Variables and Stochastic Processes". McGraw-Hill, New York.
Pearson, J.E., Bridges, W.B., Hansen, S., Nussmeier, T.A., and Pedinoff, M.E. (1976). *Appl. Opt.* **15**, 611–621.
Peck, E.R., and Obetz, S.W. (1953). *J. Opt. Soc. Am.* **43**, 505–509.
Pederson, H.M., Lökberg, O.J., and Förre, B.M. (1974). *Opt. Commun.* **12**, 421–426.
Peek, Th. H., Bolwijn, P.T., and Alkemade, C. Th. J. (1967). *Am. J. Phys.* **35**, 820–831.
Peřina, J. (1971). "Coherence of Light". Van Nostrand Reinhold, New York.
Peterson, O.G., Tuccio, S.A., and Snavely, B.B. (1970). *Appl. Phys. Lett.* **17**, 245–247.
Petley, B.W. (1983). *Nature* **303**, 373–376.
Philbert, M. (1958). *Rev. Opt.* **37**, 598–608.
Post, E.J. (1967). *Rev. Mod. Phys.* **39**, 475–493.
Powell, R.L., and Stetson, K.A. (1965). *J. Opt. Soc. Am.* **55**, 1593–1598.
Pryputniewicz, R.J. (1978). *Appl. Opt.* **17**, 3613–3618.
Pryputniewicz, R.J. (1980). *In* "Technical Digest on Hologram Interferometry and Speckle Metrology", pp. MB1-1–MB1-7. The Optical Society of America, Washington.
Pryputniewicz, R.J., and Bowley, W.W. (1978). *Appl. Opt.* **17**, 1748–1756.
Pryputniewicz, R.J. and Stetson, K.A. (1976). *Appl. Opt.* **15**, 725–728.
Pryputniewicz, R.J., and Stetson, K.A. (1980). *Appl. Opt.* **19**, 2201–2205.
Quist, T.M., Rediker, R.H., Keyes, R.J., Krag, W.E., Lax, B., McWhorter, A.L., and Zeigler, H.J. (1962). *Appl. Phys. Lett.* **1**, 91–92.
Raine, K.W., and Downs, M.J. (1978). *Opt. Acta.* **25**, 549–558.
Ramsay, J.V. (1966). *Appl. Opt.* **5**, 1297–1301.
Ramsay, J.V. (1969). *Appl. Opt.* **8**, 569–574.
Ramsay, J.V., Kobler, H., and Mugridge, E.G.V. (1970). *Solar Phys.* **12**, 492–501.
Rashleigh, S.C. (1981). *Opt Lett.* **6**, 19–21.
Reed, I.S. (1962). *IEEE Trans. Information Theory* **IT-8**, 194–195.
Rigden, J.D., and Gordon, E.I. (1962). *Proc. Inst. Radio Engrs.* **50**, 2367–2368.
Rimmer, M.P., King, D.M., and Fox, D.G. (1972). *Appl. Opt.* **11**, 2790–2796.
Rimmer, M.P., and Wyant, J.C. (1975). *Appl. Opt.* **14**, 142–150.
Ripper, J.E., Dyment, J.C., D'Asaro, L.A., and Paoli, T.L. (1971). *Appl. Phys. Lett.* **18**, 155–157.
Roberts, R.B. (1975). *J. Phys. E: Sci. Instrum.* **8**, 600–602.
Roddier, F. (1979). *In* "Higher Angular Resolution Stellar Interferometry" (J. Davis and W.J. Tango, eds); I.A.U. Colloquium No. 50, pp. 32.1–32.15. Chatterton Astronomy Department, University of Sydney.
Roddier, F., Roddier, C., and Demarcq, J. (1978). *J. Optics (Paris)* **9**, 145–149.
Roland, J.J., and Agrawal, G.P. (1981). *Opt. and Laser Tech.* **13**, 239–244.
Ronchi, V. (1964). *Appl Opt.* **3**, 437–451.
Rosen, L. (1967). *Proc. IEEE.* **55**, 118.
Ross, I.N., and Singh, S. (1983). *J. Phys. E: Sci Instrum,* **16**, 745–746.
Rotz, F.B., and Friesem, A.A. (1966). *Appl. Phys. Lett.* **8**, 146–148.
Rowley, W.R.C., and Wilson, D.C. (1972). *Appl. Opt.* **11**, 475–476.
Salimbeni, R., and Pole, R.V. (1980). *Opt. Lett.* **5**, 39–41.
Sandercock, J.R. (1970). *Opt. Commun.* **2**, 73–76.
Sandercock, J.R. (1976). *J. Phys. E: Sci. Instrum.* **9**, 566–568.
Saunders, J.B. (1955). *J. Opt. Soc. Am.* **45**, 133.
Schäfer, F.P. (1973). "Dye Lasers". Springer-Verlag, Berlin.
Schaham, M. (1982). *Proc. SPIE* **306**, 183–191.
Schawlow, A.L., and Townes, C.H. (1958). *Phys. Rev.* **112**, 1940–1949.

Schmahl, G., and Rudolph, D. (1976). *In* "Progress in Optics" (E. Wolf, ed.), Vol. XIV, pp. 196–244. North-Holland, Amsterdam.

Schmidt, H., Salzmann, H., Strowald, H., (1975). *Appl. Opt.* **14,** 2250–2251.

Schulz, G., and Schwider, J. (1967). *Appl. Opt.* **6,** 1077–1084.

Schulz, G., and Schwider, J. (1976). *In* "Progress in Optics" (E. Wolf, ed.), Vol. XIII, pp. 95–167. North-Holland. Amsterdam.

Scott, R.M. (1969). *Appl. Opt.* **8,** 531–537.

Seeley, J.S. (1964). *J. Opt. Soc. Amer.* **54,** 342–346.

Shack, R.V. and Hopkins, G.W. (1979). *Opt. Eng* **18,** 226–228.

Shankland, R.S. (1973). *Appl. Opt.* **12,** 2280–2287.

Shankoff, T.A. (1968). *Appl. Opt.* **7,** 2101–2105.

Sheem, S.K., and Giallorenzi, T.G. (1979). *Opt. Lett.* **4,** 29–31.

Shibayama, K., and Uchiyama, H. (1971). *Appl. Opt.* **10,** 2150–2154.

Shurcliff, W.A. (1962). "Polarized Light: Production and Use". Harvard University Press, Cambridge.

Shurcliff, W.A., and Ballard, S.S. (1964). "Polarized Light". Van Nostrand, Princeton.

Slepian, D., and Pollak, H.O. (1961). *Bell Syst. Tech. J.* **40,** 43–63.

Slettemoen, G.A. (1977). *Opt. Commun.* **23,** 213–216.

Slettemoen, G.A. (1979). *Opt. Acta* **26,** 313–327.

Slettemoen, G.A. (1980). *Appl. Opt.* **19,** 616–623.

Smartt, R.N. (1974). *Appl. Opt.* **13,** 1093–1099.

Smartt, R.N., and Steel, W.H. (1975). *Japan. J. Appl. Phys.* **14, Suppl 14-1,** 351–356.

Smith, H.M. (1977). "Holographic Recording Materials". Springer-Verlag, Berlin.

Smith, P.W. (1965). *IEEE J. Quant. Elect.* **QE-1,** 343–348.

Snyder, J.J. (1977). *In* "Laser Spectroscopy III" (J.L. Hall and J.L. Carsten, eds), pp. 419–420. Springer-Verlag, Berlin.

Snyder, J.J. (1981). *Proc. SPIE* **288,** 258–262.

Šolc, I. (1965). *J. Opt. Soc. Am.* **55,** 621–625.

Sollid, J.E. (1969). *Appl. Opt.* **8,** 1587–1595.

Sommargren, G.E., and Thompson, B.J. (1973). *Appl. Opt.* **12,** 2130–2138.

Steel, W.H. (1962). *Opt. Acta* **9,** 111–119.

Steel, W.H. (1964a). *Opt. Acta* **11,** 9–19.

Steel, W.H. (1964b). *Opt. Acta* **11,** 211–217.

Steel, W.H. (1965). *In* "Progress in Optics" (E. Wolf, ed.), Vol. V, pp. 147–194. North-Holland, Amsterdam.

Steel, W.H. (1967). "Interferometry". Cambridge University Press, Cambridge. Revised edition, 1983.

Steel, W.H. (1970). *Opt. Acta* **17,** 873–881.

Steel, W.H. (1975). *Opt. Commun.* **14,** 108–109.

Steel, W.H., Smartt, R.N., and Giovanelli, R.G. (1961). *Austral. J. Phys.* **14,** 201–211.

Stetson, K.A. (1969). *Optik* **29,** 386–400.

Stetson, K.A. (1970a). *Optik* **31,** 576–591.

Stetson, K.A. (1970b). *Opt. Laser Technol.* **2,** 179–181.

Stetson, K.A. (1974). *J. Opt. Soc. Am.* **64,** 1–10.

Stetson, K.A. (1975a). *Appl. Opt.* **14,** 272–273.

Stetson, K.A. (1975b). *Appl. Opt.* **14,** 2256–2259.

Stetson, K.A. (1978). *Exp. Mech.* **18,** 67–73.

Stetson, K.A. (1979). *J. Opt. Soc. Am.* **69,** 1705–1710.

Stetson, K.A., and Powell, R.L. (1965). *J. Opt. Soc. Am.* **55,** 1694–1695.

Streifer, W., Burnham, R.D., and Scifres D.R. (1977). *IEEE J. Quantum Electron.* **QE-13,** 403–404.

Stroke, G.W., and Restrick, R.C. (1965). *Appl. Phys. Lett.* **7,** 229–230.

Strong, J. (1936). *J. Opt. Soc. Am.* **26,** 73–74.

Strong, J.D., and Vanasse, G. (1958). *J. Phys. Radium.* **19,** 192–196.

Strong, J., and Vanasse, G.A. (1959). *J. Opt. Soc. Am.* **49,** 844–850.

Strong, J., and Vanasse, G.A. (1960). *J. Opt. Soc. Am.* **50,** 113–118.

Stumpf, K.D. (1979). *Opt. Eng* **18,** 648–653.

Sudol, R., and Thompson, B.J. (1979). *Opt. Commun.* **31,** 105–110.

Svelto, O. (1982). "Principles of Lasers". Plenum Press, New York.

Sweeny, D.W., and Vest, C.M. (1973). *Appl. Opt.* **12,** 2649–2664.

Takai, N., Yamada, M., and Idogawa, T. (1976). *Opt. Laser Tech.* **8,** 21–23.

Tango, W.J., and Twiss, R.Q. (1980). *In* "Progress in Optics" (E. Wolf, ed.), Vol. XVII, pp. 241–277. North-Holland, Amsterdam.

Tanner, L.H. (1966). *J. Sci. Instrum.* **43,** 878–886.

Tanner, L.H. (1967). *J. Sci. Instrum.* **44,** 1015–1017.

Terrien, J. (1958). *J. Phys. Radium.* **19,** 390–396.

Terrien, J. (1976). *Rep. Progr. Phys.* **39,** 1067–1108.

Texereau, J. (1963). *Appl. Opt.* **2,** 23–30.

Thompson, B.J., and Wolf, E. (1957). *J. Opt. Soc. Am.* **47,** 895–902.

Thompson, G.H.B. (1980). "Physics of Semiconductor Laser Devices". John Wiley, Chichester.

Tilford, C.R. (1977). *Appl. Opt.* **16,** 1857–1860.

Timmermans, C.J., Schellekens, P.H.J., and Schram, D.C. (1978). *J. Phys. E: Sci. Instrum.* **11,** 1023–1026.

Tiziani, H. (1971). *Opt. Acta* **18,** 891–902.

Tiziani, H. (1972). *Opt. Commun.* **5,** 271–276.

Tolansky, S. (1945). *Phil. Mag.* **36,** 225–236.

Tolansky, S. (1955). "An Introduction to Interferometry". Longmans, London.

Tolansky, S. (1961). "Surface Microtopography". Longmans, London.

Troup, G.J. and Turner, R.G. (1974). *Rep. Progr. Phys.* **37,** 771–816.

Tsuruta, T. (1963). *J. Opt. Soc. Amer.* **53,** 1156–1161.

Tsuruta, T., Shiotake, N., and Itoh, Y., (1968). *Japan J. Appl. Phys.* **7,** 1092–1100.

Tsuruta, T., Shiotake, N., Tsujiuchi, J., and Matsuda, K. (1967). *Japan J. Appl. Phys.* **6,** 661–662.

Turner, A.F., and Baumeister, P.W. (1966). *Appl. Opt.* **5,** 69–76.

Twyman, F. (1957). "Prism and Lens Making". Hilger and Watts, London.

Umeda, N., Tsukiji, M., and Takasaki, H. (1980). *Appl. Opt.* **19,** 442–450.

Upatnieks, J., and Leonard, C. (1970). *J. Opt. Soc. Am.* **60,** 297–305.

Urbach, J.C., and Meier, R.W. (1966). *Appl. Opt.* **5,** 666–667.

Vacher, R., Sussner, H., and Schickfus, M.v. (1980). *Rev. Sci. Instrum.* **51,** 288–291.

Vali, V., and Shorthill, R.W. (1976). *Appl. Opt.* **15,** 1099–1100.

Vanasse, G.A., and Sakai, H. (1967). *In* "Progress in Optics" (E. Wolf, ed.), Vol. VI, pp. 261–330. North-Holland, Amsterdam.

van Cittert, P.H. (1934). *Physica* **1,** 201–210.

van Cittert, P.H. (1939). *Physica* **6,** 1129–1138.

Vest, C.M. (1979). "Holographic Interferometry". John Wiley, New York.

Vittoz, B. (1956). *Rev. Opt.* **35,** 253–291, 468–480.

von Bally, G. (1979). "Holography in Medicine and Biology". Springer-Verlag, Berlin.

von der Heide, K. (1978). *Astron. Astrophys.* **70,** 777–784.

Voumard, C. (1977). *Opt. Lett.* **1,** 61–63.

Wallard, A.J. (1972). *J. Phys. E: Sci. Instrum.* **5**, 926-930.
Wallard, A.J. (1973). *J. Phys. E: Sci. Instrum.* **6**, 793-807.
Walles, S. (1969). *Arkiv för Fysik* **40**, 299-403.
Weigelt, G.P. (1975). *Optik* **43**, 111-128.
Weigelt, G.P. (1977). *Opt. Commun.* **21**, 55-59.
Weigelt, G.P. (1978). *Appl. Opt.* **17**, 2660-2662.
Weigelt, G.P. (1982). *Proc. SPIE* **332**, 284-291.
Weigelt, G., and Wirnitzer, B. (1983). *Opt. Lett.* **8**, 389-391.
Weinberg, F.J. (1963). "Optics of Flames". Butterworths. London.
Weinstein, W. (1954). *Vacuum* **4**, 3-19.
White, A.D., and Rigden, J.D. (1962). *Proc IRE (Correspondence)* **50**, 1697.
Whitford, B.G. (1979). *Opt. Commun.* **31**, 363-365.
Winter, J.G. (1984). *Opt. Acta* **31**, 823-830.
Wolf, E. (1954). *Proc. Roy. Soc.* **A225**, 96-111.
Wolf, E. (1955). *Proc. Roy. Soc.* **A230**, 246-265.
Wolfke, M. (1920). *Phys. Zeitschr.* **21**, 495-497.
Womack, K.H., Jonas, J.A., Koliopoulos, C., Underwood, K.L., Wyant, J.C., Loomis, J.S., and Hayslett, C.R. (1979). *Proc. SPIE* **192**, 134-139.
Wood, T.H. (1978). *Rev. Sci. Instrum.* **49**, 790-793.
Wyant, J.C. (1973). *Appl. Opt.* **12**, 2057-2060.
Wyant, J.C. (1975). *Appl. Opt.* **14**, 2622-2626.
Wyant, J.C. (1976). *Opt. Commun.* **19**, 120-121.
Wyant, J.C., and Bennet, V.P. (1972). *Appl. Opt.* **11**, 2833-2839.
Wyant, J.C., and O'Neill, P.K. (1974). *Appl. Opt.* **13**, 2762-2765.
Yamada, M., Ikeshima, H., and Takahashi, Y. (1980). *Rev. Sci. Instrum.* **51**, 431-434.
Yarborough, J.M., and Hobart, J. (1973). *IEEE/OSA Conference on Laser Engineering and Applications, Washington D.C.* (Post-deadline paper).
Yariv, A., and Winsor, H. (1980). *Opt. Lett.* **5**, 87-89.
Yaroslavskii, L.P., and Merzlyakov, N.S. (1980). "Methods of Digital Holography". Consultants Bureau, Plenum Publishing Company, New York.
Yokozeki, S., and Suzuki, T. (1971). *Appl. Opt.* **10**, 1575-1580.
Yoshihara, K. (1968). *Japan J. Appl. Phys.* **7**, 529-535.
Zambuto, M.H., and Fischer, W.K. (1973). *Appl. Opt.* **12**, 1651-1655.
Zelenka, J.S., and Varner, J.R. (1968). *Appl. Opt.* **7**, 2107-2110.
Zelenka, J.S. and Varner, J.R. (1969). *Appl. Opt.* **8**, 1431-1434.
Zernike, F. (1938). *Physica,* **5**, 785-795.
Zernike, F. (1950). *J. Opt. Soc. Am.* **40**, 326-328.
Zhou, W. (1984). *Opt. Commun.* **49**, 83-85.

Author index

Subject index

5 6 7 8 9 0 1 2 3 4
A B C D E F G H I J